The ESRI® Guide to

GIS

Analysis

Volume 2: Spatial Measurements & Statistics

Andy Mitchell

ESRI PRESS
REDLANDS, CALIFORNIA

ESRI Press, 380 New York Street, Redlands, California 92373-8100

Copyright © 2005 ESRI

All rights reserved. First edition 2005
10 09 08 07 06 05   2 3 4 5 6 7 8 9 10

Printed in the United States of America

Library of Congress Control Number: 2004400242

Ask for ESRI Press titles at your local bookstore or order by calling 1-800-447-9778. You can also shop online at www.esri.com/esripress. Outside the United States, contact your local ESRI distributor.

ESRI Press titles are distributed to the trade by the following:

In North America:
Ingram Publisher Services
Toll-free telephone: (800) 648-3104
Toll-free fax: (800) 838-1149
E-mail: customerservice@ingrampublisherservices.com

In the United Kingdom, Europe, and the Middle East:
Transatlantic Publishers Group Ltd.
Telephone: 44 20 7373 2515
Fax: 44 20 7244 1018
E-mail: richard@tpgltd.co.uk

Cover design by Amaree Israngkura Na Ayudhya
Interior design by Andy Mitchell

Contents

Preface vii

Acknowledgments ix

CHAPTER 1: INTRODUCING SPATIAL MEASUREMENTS AND STATISTICS 1

    What are spatial measurements and statistics? 2

    Geographic analysis with statistics 6

UNDERSTANDING DATA DISTRIBUTIONS 13

CHAPTER 2: MEASURING GEOGRAPHIC DISTRIBUTIONS 21

    Why measure geographic distributions? 22

    Finding the center 26

    Measuring the compactness of the distribution 39

    Measuring orientation and direction 45

    References and further reading 61

TESTING STATISTICAL SIGNIFICANCE 63

CHAPTER 3: IDENTIFYING PATTERNS 71

    Why identify geographic patterns? 72

    Using statistics to identify patterns 75

    Measuring the pattern of feature locations 80

    Measuring the spatial pattern of feature values 104

    References and further reading 133

DEFINING SPATIAL NEIGHBORHOODS AND WEIGHTS 135

CHAPTER 4: IDENTIFYING CLUSTERS 147

    Why identify spatial clusters? 148

    Using statistics to identify clusters 149

    Finding clusters of features 152

    Finding clusters of similar values 163

    References and further reading 181

USING STATISTICS WITH GEOGRAPHIC DATA **183**

CHAPTER 5: ANALYZING GEOGRAPHIC RELATIONSHIPS **191**

Why analyze geographic relationships? 192

Using statistics to analyze relationships 196

Identifying geographic relationships 203

Analyzing geographic processes 210

References and further reading 227

Data credits 229

Software credits 231

Index 235

Preface

In the preface to the first volume of *The ESRI® Guide to GIS Analysis,*
I wrote, "Spatial analysis has often seemed inaccessible to many GIS
users—too mathematical to understand, too difficult to implement, and
lacking in good textbooks and guides." Volume 1 seemed to me to be
exactly what was needed by GIS users without a strong background in
mathematics and statistics—a well-illustrated, accessibly written discussion
of the main methods and how they can be used to answer important questions. I noted that "ESRI plans to follow and build on this with a second
more advanced book, which will cover some of the more complex methods." But I had serious doubts about that second project, since it would
have to venture into more difficult territory, including the forbidding top-
ics of statistical inference and hypothesis testing.

As an instructor I have had abundant first-hand experience of the difficul-
ties students often have with these concepts. But I also know how power-
ful GIS can be. Ideas that used to sound impossibly dry and uninteresting
when presented on a blackboard with chalk come alive and are com-
pelling when introduced through the colorful, practical medium of GIS.
Arguments made in words are never as accessible as arguments presented
in pictures, particularly when those pictures refer to real issues, such as
public health, crime, or the environment.

Like its predecessor, this new book is a triumph. It combines the relaxed,
intuitive style of Andy Mitchell's writing and design with access to the
wealth of applications and examples that ESRI has been storing up over
the 35 years of its existence. It doesn't short-change the reader, but instead
confronts sampling, spatial dependence, and statistical inference head-
on. But it does it in a gentle way that minimizes mathematical notation,
and relies instead on an abundance of colorful graphics and interesting
examples. Many of the issues are at the cutting edge and far from settled,
including the troublesome topic of cluster detection, but readers will find
them treated in a straightforward way with plenty of directions for further,
deeper reading.

The book should be required reading for everyone who ventures into the
world of spatial analysis with GIS. The two books together cover much of
the ground, but they leave plenty of room for additional volumes, and I
for one am looking forward to ESRI's future efforts.

Michael F. Goodchild
National Center for Geographic Information and Analysis
University of California, Santa Barbara

Acknowledgments

Many people contributed their knowledge to this book. Foremost among these is Dr. Lauren Scott of ESRI, who reviewed several versions of the manuscript and provided both source material and technical guidance. The book would not have been written without her collaboration. Others, mainly from academia, contributed through their publications, which are listed at the end of each chapter.

Dr. Michael Goodchild of the University of California, Santa Barbara, reviewed several versions of the manuscript and provided valuable comments on the direction of the book, and also wrote the preface. Dr. Arthur Getis and Jared Aldstadt of San Diego State University reviewed sections of the text for technical accuracy. Dr. Arthur Lembo Jr. of Cornell University, Dr. Thomas Balstrøm of the University of Copenhagen, and Dr. Mak Kaboudan of the University of Redlands also reviewed portions of the manuscript and provided valuable comments. Several people at ESRI reviewed the manuscript, including Clint Brown, Witold Fraczek, Steve Kopp, Steve Lynch, Mike Minami, and Mark Smith.

A number of organizations provided the data used to create the map examples in the book. They are listed in the "Data credits" section. Several people at ESRI—including Hugh Keegan, Mark Smith, Lee Johnston, John Calkins, and Damian Spangrud—also made data sets available. Nathan Warmerdam assisted with creating examples.

Many at ESRI Press and other departments at ESRI helped with the production of the book. Robin Rowe was the project manager. Michael Karman edited the book. Denise Marshall revised the cover art, Judy Hawkins managed the publication process, Michael Hyatt assisted with production, and Cliff Crabbe coordinated printing.

Finally, once again, special thanks to Jack Dangermond and Clint Brown, who recognized the value of publishing a guide to GIS analysis, and provided the support for writing it.

1

Introducing spatial measurements and statistics

Spatial measurements and statistics allow you to quantify patterns and relationships. That makes it easier to compare sets of features and to track changes over time. You can also calculate a probability that a pattern or relationship actually exists.

In this chapter:

- *What are spatial measurements and statistics?*
- *Geographic analysis with statistics*

GIS analysis is about getting answers to questions so you can make intelligent decisions. The previous book in this series showed you how to do GIS analysis with maps. In some cases, the map was the analysis. In other cases, you used GIS tools and methods to create new data that was then displayed on a map so you could analyze it and draw conclusions.

Sometimes, making a map may be enough to get the answers you need. But trying to draw conclusions from a map isn't always easy. How you classify and symbolize features and values on a map can obscure the information, and humans see patterns and relationships everywhere— even sometimes when they don't really exist.

Over the past 50 years or so, geographers, regional scientists, ecologists, economists, and others have developed tools to describe the distribution of a set of features, to discern patterns, and to measure relationships between features.

These tools rely on statistics to cut through the map display and get right at the patterns and relationships in the data. Space is a fundamental component of these statistics. That's what sets them apart from traditional statistics used to analyze aspatial data (tables of data values). The locations of the features and in many cases the spatial relationship between them (distance, for example) are considered, along with the attribute values associated with the features. If you just tried to analyze the attribute values by themselves, using traditional statistics, you'd get a false picture of what's occurring.

What if you could find the center of a group of influenza cases without guessing? Or clearly see the overall direction of a set of storm tracks? What if the GIS could identify clusters of burglaries for you?

Spatial statistics tools can help you perform these tasks, and others—tasks you may already be doing with maps. But spatial statistics open up a new set of questions you could be asking, to get even better information and be even more confident in your decisions: How sure am I that the pattern I'm seeing isn't simply due to a random occurrence? To what extent does the value of a feature depend on the values of surrounding features? How well does the value of one attribute predict the value of another? What are the trends in the data?

Statistics describe or summarize large amounts of data, useful in geographic analysis where you're often dealing with large datasets. Having a summary statistic—such as the center of features or the directional trend—makes it easy to compare sets of features or track changes over time without having to guess.

Statistics also let you derive information from a sample of features, and apply your conclusions to the whole set of features in your study area. If a sample of a plant species creates a clustered pattern, you can conclude that the species generally appears in clusters.

Statistics help you predict unknown values from known sample values. If you establish a relationship between feature values, you can predict where certain other values will occur. Knowing that landslides have occurred on slopes of a certain angle, soil moisture, and vegetation cover lets you find other slopes with these values and zone them as susceptible to landslides. The query capability of GIS—finding areas that match a set of criteria— lets you put the predictions to work.

Maybe most importantly, statistics let you verify your conclusions. You can assign a probability that your conclusions are true, and thus know how confident you can be in the decisions you make.

So why haven't people been using statistics for geographic analysis all along? One reason is that statistics are, after all, statistics—they're perceived as hard to understand and to use. Another is that statistical tools haven't been available in commercial GIS software. Many of the statistical tools have been used primarily in academic research, or limited to use in a specific discipline. People had to write their own software routines to perform their analyses. Recently, though, spatial statistics packages—such as CrimeStat® and SpaceStat™—have begun to appear. Spatial statistics tools are also appearing in comprehensive statistics software such as SAS® and S+®, and in commercial GIS, including ArcGIS® 9.

Because of the heretofore limited availability of these tools, most GIS users have not been aware of them and how they can be applied. That's where this book comes in. We want to introduce you to the most commonly used spatial statistical tools—and those most helpful to GIS users— and show how they can be applied in a range of disciplines, from crime analysis to habitat conservation. The ultimate goal of this book is to help you extract information that is already in your GIS database (in which you've undoubtedly already invested substantial amounts of time and money), but that might not be obvious simply by creating a map.

In this book we've identified some common questions that spatial statistics can answer.

How are the features distributed?

Statistics can describe the characteristics of a set of features, including the center of the features, the extent to which features are clustered or dispersed around the center, and any directional trend. Analyzing the distribution of features is useful for studying change over time—for example, to see where the center of cases of a particular disease moves over the course of several months—or for comparing two or more sets of features.

What is the pattern created by the features?

You can use statistics to measure whether—and to what extent—the distribution of features creates a pattern. If you find that cases of a disease form a clustered pattern, there are likely local sources (perhaps ponds harboring infected mosquitoes).

You can also identify patterns in the distribution of attribute values associated with the features. For example, you might calculate the degree to which student test scores in a city are clustered. If attendance areas with similarly high or low scores occur together, it may mean money and other resources are not being distributed equally.

Where are the clusters?

Finding individual clusters is useful when you need to take immediate action or when you want to find the cause of the cluster, so you know what action to take. A public health department would take immediate action to notify people living where a flu cluster has been identified to watch for symptoms. They could then try to identify the source of the outbreak—if it's a school, they would know to begin inoculating the children.

You can also use statistics to identify clusters of features with similar values. A tax assessor could create neighborhoods by identifying clusters of block groups with similar median house values.

What are the relationships between sets of features or values?

While the first three questions focus on the distribution of features in a single layer, this question deals with the relationships between two or more layers. You can determine if the features—or values associated with the features—occur together, and measure the strength of the relationship. A public health analyst could determine the extent to which economic or demographic factors and the quality of infant health are related in neighborhoods across a county. Once you've identified a relationship, you can predict where features or particular attribute values will occur.

The book assumes you have little or no knowledge of statistics, but some familiarity with GIS. Four sections that deal with general statistical concepts applicable throughout the book appear between chapters: "Understanding data distributions," "Testing statistical significance," "Defining spatial neighborhoods and weights," and "Using statistics with geographic data."

The emphasis in this book is on applying the statistical tools to get meaningful results, rather than on the mathematical theory behind the statistics. However, enough background and context is presented to understand the concepts behind the tools.

There are many spatial statistical tools and methods available, more than are discussed in this book. The ones included are those that are widely used and applicable to GIS analysis across a range of disciplines. Researchers continue to improve existing tools, and develop new ones, to better capture how geographic phenomena behave. The tools presented here represent current published versions. The references at the end of each chapter contain additional information on these tools, as well as others you might find useful.

A couple of related fields beyond the scope of this book are also worth exploring. One involves predicting values in spatially continuous data from a set of sample points (a field known as geostatistics). Geostatistics has primarily been used to study air pollution and soil contamination, and to explore for oil and natural gas, but has many other applications. Another related field involves measuring the shape and form of individual features—for example, by comparing areal extent to length of boundary for an area feature, such as a patch of forest. Measures of shape and form have often been used in landscape ecology and biogeography to study potential wildlife habitat areas and corridors.

Another related area is spatial modeling, encompassing everything from suitability models you can build in a GIS, to mathematical models that predict the behavior of fires and floods, to research models that predict the behavior of people or animals. The methods discussed in this book are often incorporated into spatial models.

Geographic analysis with statistics uses mathematical equations to draw conclusions about the characteristics, patterns, and relationships of geographic data. The process is similar to the statistical analysis you'd do with aspatial data, although using statistics with geographic data entails additional considerations.

Frame the question

You start an analysis by figuring out what information you're trying to get. In descriptive statistics this usually takes the form of a question: Where is the center of crimes? What is the overall direction of the storm tracks? In inferential statistics, the analysis is stated as a hypothesis: Burglaries are more clustered than auto thefts. Landslides in this area tend to occur more frequently on slopes over 30%. To ensure impartiality, statisticians structure the analysis assuming that the inverse of the hypothesis is true—burglaries are not more clustered than auto thefts; landslides are equally likely to occur on any type of slope. They then set out to decide whether to reject this null hypothesis, or not. (See "Testing statistical significance.")

Understand your data

In general, you can analyze features using location alone, or using location influenced by an attribute value. The geographic representation of the data and the type of attribute values will influence the statistical methods you use.

Geographic features are either discrete or spatially continuous.

Discrete features can be points, lines, or areas. Points are used to represent either stationary features (stores and pollution monitoring stations) or events that occur at a specific place and time (burglaries and earthquake epicenters). Lines can be disjunct (elk migration routes) or connected in a network. Discrete areas are usually distinct and separate, but may share a border or even overlap as fire boundaries often do.

Burglaries (points), bighorn sheep migration routes (lines), and bobcat habitat (areas) are examples of discrete features.

Spatially continuous features—temperature and precipitation are oft-cited examples—are found and can be measured anywhere and everywhere. This type of data is also referred to as a continuous field, and is usually represented as a surface.

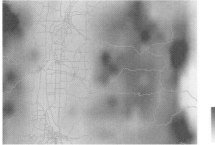

High

Low

Average annual rainfall (shown with roads) is an example
of spatially continuous data represented as a surface.

Spatially continuous categorical data, such as land-cover types, is represented as contiguous areas defined by boundaries.

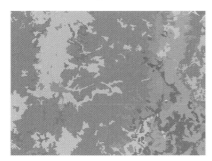

Land-cover categories are represented
using contiguous areas.

Summarized data is also represented by contiguous areas. Rather than representing what's occurring at any given location inside the area, though, the attributes associated with the areas are summaries of what's inside them—the number of senior citizens in each census tract, or the percentage of harvestable timber in each watershed. The value applies to the entire area, not any specific location within it.

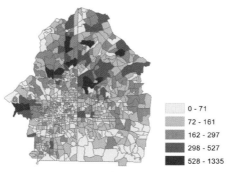

	0 - 71
	72 - 161
	162 - 297
	298 - 527
	528 - 1335

Census block groups color coded by the number of senior citizens in each is an example of summarized data.

Attribute values include nominal, ordinal, interval, and ratio data.

Nominal (categorical) data describes features of a similar type. For example, you can categorize parcels by their land use, or crimes by whether they're burglaries, assaults, thefts, and so on. You might look for the center of all crimes (an entire layer) or the center of burglaries (a subset).

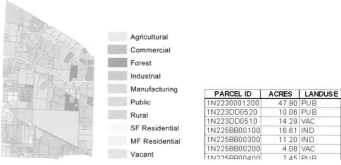

	Agricultural
	Commercial
	Forest
	Industrial
	Manufacturing
	Public
	Rural
	SF Residential
	MF Residential
	Vacant

PARCEL ID	ACRES	LANDUSE
1N2230001200	47.90	PUB
1N223DD0520	10.06	PUB
1N223DD0510	14.29	VAC
1N225BB00100	16.61	IND
1N225BB00300	11.20	IND
1N225BB00200	4.08	VAC
1N225BB00400	7.45	PUB

Parcels color coded by land-use categories

Ordinal (ranked) data describes data that is ordered from high to low. You only know where a feature falls in the order—you don't know how much higher or lower a value is than another value. For example, you know that a city with a livability rank of 3 is lower than one ranked 2 and higher than a 4, but you don't know how much lower or higher.

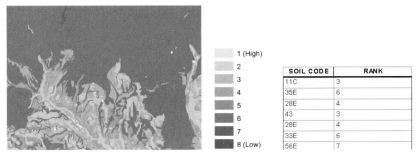

	1 (High)
	2
	3
	4
	5
	6
	7
	8 (Low)

SOIL CODE	RANK
11C	3
35E	6
28E	4
43	3
28E	4
33E	6
56E	7

Soil types ranked by suitability for agriculture

Interval data (quantities), on the other hand, does tell you relative magnitude—you know that a house with a value of $400,000 is worth twice as much as one with a value of $200,000. Interval data can be the number of something—the number of employees at each business, the number of twelfth graders in each attendance area—or a value representing a magnitude—soil pH, store revenue.

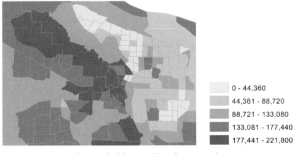

	0 - 44,360
	44,361 - 88,720
	88,721 - 133,080
	133,081 - 177,440
	177,441 - 221,800

FIPS	MEDIAN VALUE
41051004300	66600
41051004101	37800
41051007000	143500
41051004200	36400
41051004001	38600

Census tracts color coded by median house value

Ratios show you the relationship between two quantities, and are created by dividing one quantity by another for each feature. For example, dividing the number of people in each census tract by the number of households gives you the average number of people per household in each tract. Proportions (what part of a total each value is—often represented as a percentage) and densities (the quantity per unit area) are types of ratios.

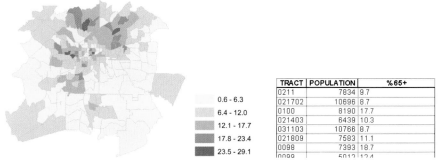

	TRACT	POPULATION	%65+
0.6 - 6.3	0211	7834	9.7
	021702	10696	8.7
6.4 - 12.0	0100	8190	17.7
	021403	6439	10.3
12.1 - 17.7	031103	10766	8.7
	021809	7583	11.1
17.8 - 23.4	0098	7393	18.7
23.5 - 29.1	0099	5012	12.4

Census tracts color coded by percentage of the population age 65 and over

Interval and ratio data are continuous values—each feature potentially has a unique value anywhere in the range between the highest and lowest values. Knowing some basic characteristics about your data, such as whether there are any extreme values (outliers) and how bunched or spread out the values are, will help you draw correct conclusions from your analysis results. (See "Understanding data distributions.")

Choose a method

There may be more than one method you can use to answer your original question or analyze your hypothesis. Your choice depends on the type of data you have, your analysis, and—in some cases—the differences between methods.

Certain types of data are appropriate for certain types of analysis. For discrete features, you can analyze the distribution of the features themselves (whether grocery stores are clustered), or the distribution of an attribute associated with the features (whether stores with high revenue are clustered).

For spatially continuous phenomena, the distribution of values associated with the phenomenon is what you're interested in—analyzing the distribution of cells in a surface or a set of contiguous areas tells you nothing. The same is true of summarized data—you're interested in the distribution of values associated with the areas; since the areas themselves cover the entire region, no information is gained by analyzing their distribution.

If you're analyzing the distribution of feature values, you'll be dealing for the most part with continuous values. When identifying patterns and clusters, you'll want to use ratios—especially when analyzing data summarized by contiguous areas—because ratios even out the differences between large and small areas. That's important if you're interested in the concentration of features or values—a large census tract may have a large number of seniors, but they may be spread out over a wide area.

Calculate the statistic
While the emphasis in this book is on the application of the statistics, we do present the equations for each statistic along with an explanation. Since the GIS software performs the calculations, you could use a statistic without knowing the math behind it; by understanding how the statistic is derived you'll be better able to decide which statistic is best for your analysis as well as avoid drawing incorrect conclusions from your results.

Some of the statistical tools require you to provide parameters. For example, you may need to specify the nature of the influence of features on each other, such as the distance within which the prices of surrounding homes influence the assessed value of a house. (See "Defining spatial neighborhoods and weights.")

Interpret the statistic
Descriptive statistics calculate a value that can be displayed on a map—the center is an x,y coordinate location; a directional trend can be displayed using an ellipse.

Other statistics calculate a number that tells you whether there is a pattern or relationship. Often the number is within a range—the position in the range indicates the nature of the pattern (whether clustered or dispersed, for example) or the relationship, and its strength. But, to really know whether the result is meaningful, you need to test its statistical significance.

Test the significance of the statistic
The null hypothesis essentially states that there is no pattern or relationship. Significance tests help you decide whether you should or should not reject the null hypothesis.

You first decide the risk you are willing to accept for being wrong. This degree of risk, often referred to as the confidence level, is expressed as a probability ranging from 0 to 1.

Statisticians accept the null hypothesis unless there is a very small chance that they would be wrong to reject it. If you're deploying police officers, knowing that you can be 80% sure (0.20 confidence level) that burglaries are clustered in a particular area may be enough to decide to send them there. If, however, you're trying to pinpoint the cause of an outbreak of disease, you'd probably want to be at least 95% sure (0.05 confidence level) that the clusters you have identified did not occur simply by chance.

Usually, the software you're using runs the appropriate test when it calculates the initial statistic. The test calculates a statistic that you compare to a critical value—based on the confidence level—to determine whether the results are significant at that confidence level. If the test statistic exceeds the critical value, you'd be right to reject the null hypothesis. The results are said to be statistically significant at that level. (See "Testing statistical significance.")

Question the results

Finally, even if the results of your analysis are statistically significant, you'll want to question them. The geographic scale you're working at, where the study area boundaries fall, the type of data you're using, the quality of the data, and how you define proximity between features all influence the results. For example, your results may be very different if you specify straight-line distance as opposed to travel time when defining how close features are to each other.

In many cases, you'll want to compare the results for the features you're analyzing to a control group. Crimes may form clusters just because those are the locations where people live; however, a cluster of crimes occurring where the population density is low may indicate a true hot spot. There are often local and regional trends in geographic data, so the outcome of your analysis may be predetermined, to some extent. (See "Using statistics with geographic data.")

The conclusions you draw from your analysis should be used in conjunction with other information, including your knowledge of the features, when making a decision. You may want to use alternate methods to confirm the results of your analysis. Statistical analysis is only one of several inputs—along with economic and political factors—to the decision-making process.

Understanding data distributions

To identify trends, patterns, and relationships, spatial measurements and statistics analyze distributions of features. Understanding the characteristics of data distributions will help you reach the correct conclusions from your analysis. Also, knowing how your data is distributed is useful before even starting the analysis—for example to spot extremely high or low values that might throw off the results of your analysis.

A type of bar chart known as a histogram shows the number of features having a particular value—the frequency distribution. For continuous values (interval and ratio data), each feature can potentially have a unique value, so ranges of values are used.

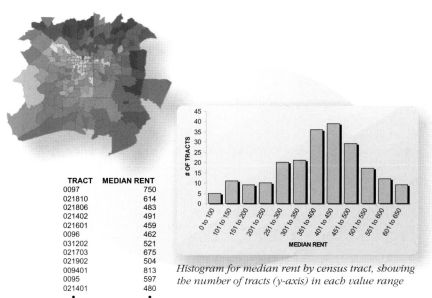

TRACT	MEDIAN RENT
0097	750
021810	614
021806	483
021402	491
021601	459
0096	462
031202	521
021703	675
021902	504
009401	813
0095	597
021401	480

Histogram for median rent by census tract, showing the number of tracts (y-axis) in each value range

From the histogram you can create a frequency curve, which eliminates the ranges and presents the values as continuous along the x-axis. You can use a spreadsheet to create a histogram or a curve. Many GIS software packages also let you create these charts.

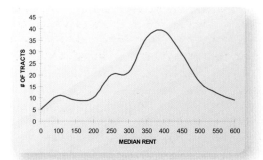

Frequency curve for median rent by census tract

Certain frequency distributions occur routinely, with a variety of data. Mathematicians have identified and described their ideal form. Two in particular are used often in spatial statistics.

The normal distribution often occurs for phenomena where the values are similar, but some are higher and some lower, such as annual rainfall over a number of years. Most years will have close to the average amount of rainfall, but there will be a few very wet years and a few very dry ones. The fre-

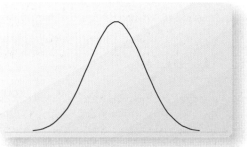

Frequency curve for a normal distribution

quency curve for a normal distribution is the classic symmetrical bell curve in which most values cluster in the center of the curve, and there are as many values on the left side of the curve as on the right. Given enough readings over a period of time, most values will be close to the mean.

The Poisson distribution, named after the French mathematician Siméon Poisson who described the distribution in the late 1830s, occurs when there are extreme events in space and time, like large-magnitude earthquakes. If the events occur randomly, there will be many time periods in which few or no events occur and very few time periods in

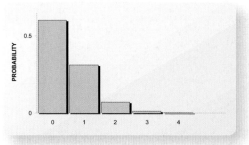

An example of a Poisson distribution, with a mean of 0.5. The y-axis shows the probability of a given number of events occurring.

which many events occur. In these cases, the mean will often be less than one, and the probability of no events occurring during the time period

will be high. In cases where the mean is greater than one (that is, on average more than one event occurs during the time period), the distribution
will look different—the probability of one or more events occurring will
be greater than the probability of none occurring.

Frequency distributions are described by measuring their characteristics.

The mean and median are
both measures of the cen-
tral value. The mean is the
average value—the attri-
bute values are summed
and divided by the number
of values. The median is
the middle value—half the
values are higher, and half
lower. In a normal distribution, the mean and median
are equal.

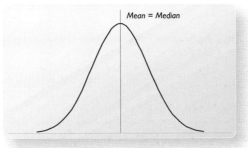

*In a normal distribution, the mean and median are
equal.*

Most geographic distributions do not fit the normal curve. In this example,
the curve is skewed toward the high values, creating a tail. The high values cause the mean to be higher than the median.

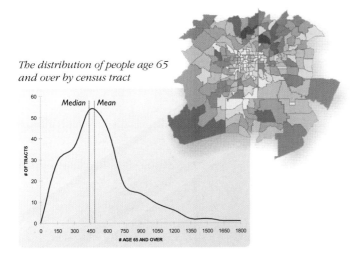

*The distribution of people age 65
and over by census tract*

Another characteristic of a distribution is the extent to which values vary from the mean, known as the variance. The larger the variance, the more dispersed the values are around the mean.

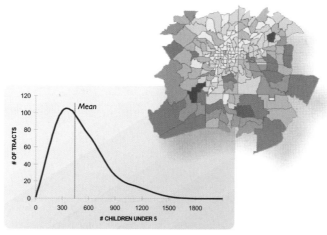

The distribution by census tract of the number of children under 5 (top map and chart) and of adults 30 to 49 years old (below). Most tracts have close to the mean number of children under 5, while the number of 30- to 49-year-olds in any tract can vary quite a bit from the mean.

The variance is calculated by subtracting the mean from each value, summing these differences, and dividing by the number of values. The difference for values less than the mean will be negative, so all the differences are squared to make sure they're positive before they're summed.

$$\sigma^2 = \frac{\sum_i (x_i - \bar{x})^2}{n}$$

The variance is calculated by subtracting the mean from each value, squaring the difference, summing all the resulting squares, and dividing by the total number of values in the set.

Since the difference from the mean is squared to calculate the variance, the values are in squared units rather than the original units. By taking the square root of the variance, the values are calculated back into the original units (feet, meters, inches, dollars, or whatever) required by some statistics. This measure is known as the standard deviation.

$$s = \sqrt{\frac{\sum_i (x_i - \bar{x})^2}{n}}$$

The standard deviation is the square root of the variance.

In a normal distribution, a certain proportion of the values will be within a certain number of standard deviations (plus or minus) of the mean.

- Plus or minus one standard deviation from the mean will contain about 68% of values.

- Plus or minus two standard deviations will contain about 95% of values.

- Plus or minus three standard deviations will contain over 99% of values.

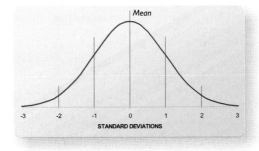

Distributions in which the mean and median are equal may still not be normal if they don't match the expected dispersion around the mean. These distributions will appear flatter or more peaked than a normal distribution.

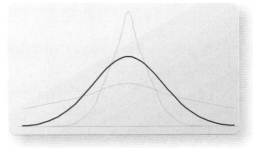

Traditional (nonspatial) statistics deal with distributions of values that describe something—the test scores of students or the revenues generated by a chain of stores. Spatial measurements and statistics deal with distributions of these descriptive values (stored as attribute values for features in a GIS database) but also with distributions of values derived from the spatial arrangement of features. If you measured the distance between each burglary and the point at the center of the burglaries, you could analyze the distribution of distance values to get a measure of how concentrated the burglaries are.

Suppose you hypothesize that assaults in a city are spread evenly and ubiquitously throughout. If you impose a grid structure on top of the city and count the actual number of assaults in each cell, you're on your way to testing this hypothesis. The histogram for this data would have

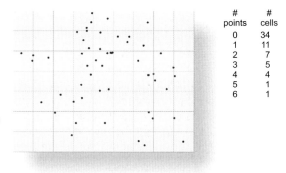

# points	# cells
0	34
1	11
2	7
3	5
4	4
5	1
6	1

the lowest count to the highest count along the x-axis, while the y-axis would show the number of cells with a given count. This distribution would be compared to the distribution had the assaults been spread evenly across the city.

Similarly, you could measure the distance between each assault and the one closest to it, and create a frequency distribution of the distances. The mean distance between nearest neighbors can be calculated and compared to what the mean distance between neighbors would be if the assaults were randomly distributed across the city.

To analyze the spatial distribution of attribute values, the frequency distribution represents the frequency of the values influenced by the distance between features. Suppose you wanted to know whether census tracts with a high percentage of an ethnic group were clustered or not. You'd weight the percentage by the distance between tracts, create the frequency distribution of these weighted values, and compare that to the distribution of the same set of values assigned randomly to the tracts.

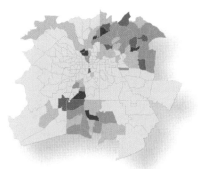

Spatial clusters are identified as nearby features that have similar values.

In the examples above, superimposing a grid and counting the number of assaults within each cell to measure the distribution starts to address the spatial nature of the data. However, the frequency distribution for two sets of features could be identical, even though their patterns are very different.

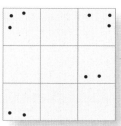 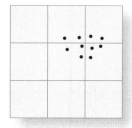

Though these distributions both have the same number of cells containing zero, two, and three points, their patterns are quite different.

Spatial statistics let you compare the spatial distribution of a set of features to a hypothetical random spatial distribution to perform true spatial pattern analysis. In most cases, the dispersion of values around the mean (variance or standard deviation) is used as the basis of comparison. To the extent that your distribution differs from a random distribution, there is a trend or a pattern in the data. See "Testing statistical significance" and "Using statistics with geographic data" for more on comparing spatial distributions.

OUTLIERS

Outliers are exceptionally high or low values, beyond what you'd expect even with a skewed distribution. Since outliers can throw off the results of your analysis, you need to know if there are any in your data. The frequency curve might give you a hint there is an outlier, but to be certain you need to sort the values in a table, or plot the individual data values on a graph.

Outliers often represent data errors. Values can be entered incorrectly in the database or associated with the wrong feature. Once you've identified any outliers, you need to check your original data sources to make sure the values for those features are correct. Missing values often show up as outliers. If you can't obtain a valid value for the feature, you may need to remove it from the dataset before performing your analysis.

Outliers are not always data errors. They may, in fact, represent valid but unexpected values, like a mansion in an otherwise modest neighborhood, which will have a much higher value than its neighbors. Or, an outlier might reflect a previously unknown condition, which alters assumptions of your analysis or changes your approach. (A block group having an unexpectedly high number of crimes, for example, might point to a previously unknown drug-trafficking hot spot.)

This parcel has a much greater value than any of the surrounding parcels.

PARCEL ID	VALUE	LANDUSE
1S203CC03000	110,370	SFR
1S203CC02900	117,480	SFR
1S203CD01300	857,340	PUB
1S203CD01200	138,410	SFR
1S203CD03100	113,360	SFR
1S203CD03000	122,190	SFR
1S203CD02900	107,580	SFR

When outliers represent valid data values, you'll want to measure the influence these outliers have on your analytical results. One way to do this is to run the analysis with and without the outliers. If the results are very close to each other, the outliers aren't having a strong impact on your results. If the results deviate strongly, you may need to find analysis methods that will not be as sensitive to the presence of any outliers.

Your data may also contain spatial outliers—features that are far from other features. As with value outliers, a spatial outlier could be a valid feature. Or it could be a feature that doesn't really exist and needs to be removed from the database. Even more likely, the outlier could represent a feature that is in the wrong place—perhaps an address that was geocoded to the wrong location.

References

Burt, James E., and Gerald M. Barber. *Elementary Statistics for Geographers*. Guilford, 1996.

Earickson, Robert J., and John M. Harlin. *Geographic Measurement and Quantitative Analysis*. Macmillan, 1994.

Ebdon, David. *Statistics in Geography*. Blackwell, 1985.

2 Measuring geographic distributions

Measuring the distribution of a set of features allows you to calculate a characteristic of the distribution, such as its center, compactness, or orientation. You can use this value to track changes in the distribution or to compare distributions of different features.

In this chapter:

- Why measure geographic distributions?
- Finding the center
- Measuring compactness
- Measuring direction and orientation

You can use GIS to calculate a statistical value representing a characteristic of a distribution of features, such as its center, the extent to which features are clustered or dispersed around the center (its compactness), or whether the features trend in a particular direction. You display—on a map—the number the statistic gives you, to see a graphic representation of the characteristic.

The center, dispersion, and directional trend of assaults occurring over the course of a year

Calculating a statistic is useful because visual analysis can be misleading—the characteristic you're interested in may not be apparent by simply mapping features.

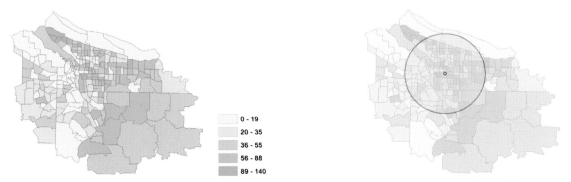

	0 - 19
	20 - 35
	36 - 55
	56 - 88
	89 - 140

American Indian population, by census tract. The map on the right shows the population center and the degree of dispersion.

If you have several features at a single location, the center of the distribution may not be obvious. This is true even if you use graduated symbols.

The center of burglaries occurring over the course of a year (orange circle). The map on the right uses graduated symbols to show the number of burglaries occurring at each location.

COMPARING THE DISTRIBUTIONS OF DIFFERENT FEATURES

A planner analyzing a regional transportation network could calculate and map the centers of manufacturing, retail, and financial activity, using the locations of businesses. If the retail and financial centers are near each other, the planner would want to consider a location between them for a bus or light-rail hub. The industrial center would become the location for a truck or rail freight hub.

The center of manufacturing, of retail, and of financial businesses in a region (from left to right)

A crime analyst might map the dispersion of auto thefts, assaults, and other thefts to see how the distributions compare. The results might point to different strategies for addressing the different types of crime.

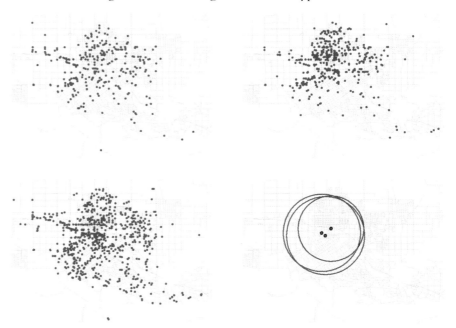

The dispersion of auto thefts (shown in blue), assaults (red), and other thefts (brown) over one year. Auto thefts and other thefts are similar (blue and brown circles), while assaults are more concentrated and centered to the northeast.

TRACKING CHANGE IN A DISTRIBUTION OF FEATURES
A crime analyst might want to see if burglaries shift between day and night in order to help commanders decide where to assign patrols at different times of day. The analyst would plot the center of daytime burglaries occurring over a period of several months and the center of nighttime burglaries over the same period.

The center of daytime burglaries (left map) is south of the freeway, while the center of nighttime burglaries is to the north.

Similarly, an epidemiologist interested in finding out where dengue fever is spreading could calculate and map the center of cases week by week. If the centers are moving toward the northeast, the epidemiologist could tell doctors where the disease is headed.

The center of dengue fever cases in a village during week two (left map), week three, and week four of an outbreak

Finding the center of a group of features is useful for tracking change in the distribution. It's also useful as a quick way to find a good location for something that has to be centrally located, such as a public facility or a store.

THREE KINDS OF CENTER

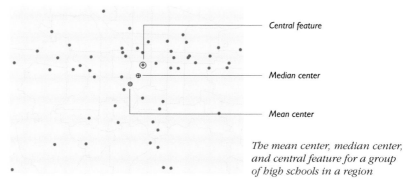

Central feature

Median center

Mean center

The mean center, median center, and central feature for a group of high schools in a region

Calculate the *mean center* for features where there is no travel to or from the center, such as with crimes or wildlife observations. For example, a wildlife biologist could calculate the mean center of elk observations within a park over several years to see where elk congregate in summer and winter, in order to tell park visitors the best places to view elk. The mean center is the average x-coordinate and y-coordinate of all the features in the study area.

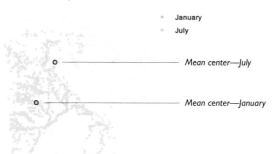

January

July

Mean center—July

Mean center—January

The mean center of elk sightings in January and in July, over four decades

Calculate the *median center* (also called the center of minimum distance) if you need to find the best location for something that needs to be centrally located. The median center is the location having the shortest total distance to all features in the study area. To find the best place in a neighborhood to park a bookmobile, you'd calculate the median center of population. The median center is usually calculated using straight-line (Euclidean) distance, although other measures of distance, such as travel time, can also be used.

	11 - 84
	85 - 164
	165 - 318
	319 - 619
	620 - 1121

The median center of population age 65 and over (by census block group)

Calculate the *central feature* if you want to find the feature that is the shortest total distance from all the other features. This represents the most centrally located feature, in terms of distance. If, for each feature, you summed the distance to all other features in the study area, the one having the lowest value (shortest total distance) would be the central feature. If you wanted to hold a workshop for high school teachers in a region, you could calculate the central feature to identify the most centrally located high school.

High schools, with the central feature circled

Center	What it represents	What it's good for
Mean	The average x-coordinate and average y-coordinate for all features in the study area	Tracking changes or comparing distributions
Median	The x,y coordinate having the shortest distance to all features in the study area	Finding the most accessible location
Central feature	The feature having the shortest total distance to all other features in the study area	Finding the most accessible feature

FINDING THE CENTER OF FEATURES BY LOCATION OR BY ATTRIBUTE

You can calculate the center of the features by location alone—which gives you an unweighted center—or as influenced by an attribute value in the feature layer's data table—which gives you a weighted center. You can calculate the unweighted or weighted mean or median center, or central feature.

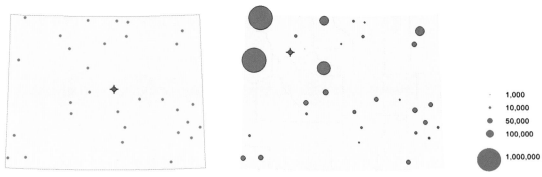

1,000
10,000
50,000
100,000
1,000,000

The mean center of state and national parks in Wyoming (left map), and the center weighted by the number of visitors per year (right map)

The unweighted center is often used for incidents or events that occur at a place and time, such as crimes.

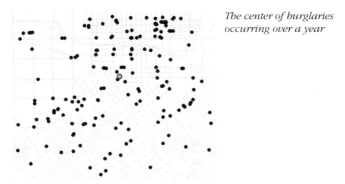

The center of burglaries occurring over a year

The weighted center is often calculated for stationary features such as stores. A retail analyst scouting warehouse locations for a grocery store chain would calculate the weighted center using store locations along with store size or revenues to ensure the warehouse is closer to stores that do more business.

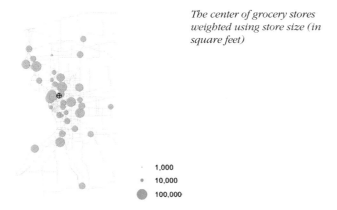

The center of grocery stores weighted using store size (in square feet)

1,000
10,000
100,000

Unlike incidents or events (such as crimes), the distribution of stationary features is usually predetermined, since they have been placed in their location for a reason. So, finding the unweighted center of stationary features may not be very meaningful. However, finding the center influenced by an attribute could yield useful information.

Contiguous areas are like other stationary features; it's not their distribution you're interested in so much as the distribution of the attribute values associated with the areas.

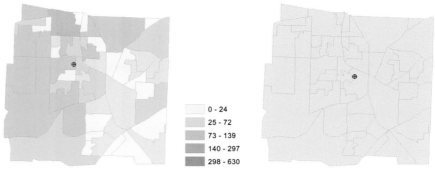

	0 - 24
	25 - 72
	73 - 139
	140 - 297
	298 - 630

The center of census block groups weighted by the Asian population (left map), and the unweighted center of the block groups. The latter provides little useful information, since the block groups are contiguous.

You can display the weighted center in 3D. For earthquakes, you might use the magnitude of the quakes as the weight.

The center of earthquakes, in 3D, weighted by magnitude and showing depth below the surface

Specifying a weight
The weight is an interval or ratio value associated with a feature. The higher the value, the greater the weight for that feature. For example, if you wanted to find the most accessible location to hold a seminar for workers in the financial sector, you'd calculate the weighted center of businesses using the number of employees as the weight value.

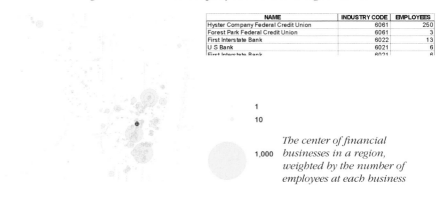

NAME	INDUSTRY CODE	EMPLOYEES
Hyster Company Federal Credit Union	6061	250
Forest Park Federal Credit Union	6061	3
First Interstate Bank	6022	13
U S Bank	6021	6
First Interstate Bank	6021	6

1
10

1,000 *The center of financial businesses in a region, weighted by the number of employees at each business*

A social service agency looking for a location for a senior center could calculate the weighted center of census block groups using the number of people age 65 and over as the weight value.

BLOCK GROUP	AGE 65+
131210016001	26
131210022002	78
131210023004	148
131210008005	101
131210018003	70
131210022005	69
131210018004	141
131210023002	74
131210023001	44

0 - 20
21 - 58
59 - 114
115 - 200
201 - 300

The center of census block groups weighted by the number of people age 65 and over in each block group

The weighted center is useful when analyzing the distribution of values associated with areas. The areas can be either contiguous or discrete. A wildlife biologist looking for a site for a research station to study a particular species could calculate the weighted mean center of potential habitat areas for that species. The weight would be the number of individuals within each territory (or the areal extent of each area as an approximation of the number of individuals).

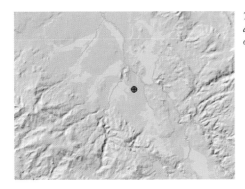

The center of bobcat habitat areas, weighted by the size of the areas

If you want to find the weighted center using the lowest values rather than the highest, you'll need to make sure the lowest value gets the highest weight. One common approach is to use the inverse of the attribute value you're using as the weight. To do this, add a new attribute to the feature data table and calculate the new values by dividing the original values into 1. To find the weighted mean center of lowest median home value for a set of census tracts—in order to find a location for a financial aid office, for example—you'd calculate a new weight value by dividing the median home value of each tract into 1. This creates a fraction—the lower the median value, the larger the fraction and hence the greater the weight.

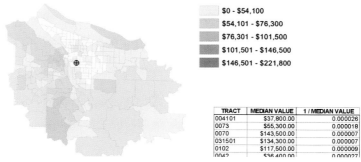

	$0 - $54,100	
	$54,101 - $76,300	
	$76,301 - $101,500	
	$101,501 - $146,500	
	$146,501 - $221,800	

TRACT	MEDIAN VALUE	1 / MEDIAN VALUE
004101	$37,800.00	0.000026
0073	$55,300.00	0.000018
0070	$143,500.00	0.000007
031501	$134,300.00	0.000007
0102	$117,500.00	0.000009
0042	$36,400.00	0.000027

The center of low median home value, obtained by calculating the inverse of the median value

With data summarized by area, you can calculate the weighted center using the total number within each area, or you can use a ratio to find the center of concentration. For example, to find a location for an office providing social services for an ethnic group you could use the population of the group in each census tract. That will weight the center toward the tracts with the greatest population. These may in fact be the largest tracts. What you may really want is the center of concentration of the ethnic group. To calculate this you'd use as the weight the value for density (population per unit area, such as square kilometer or square mile), or the percentage of the ethnic population in each tract. For density you'd calculate the ethnic population divided by the area of the tract and use this value as the weight. For the percentage, you'd divide the ethnic population by the total population of the tract and use this value as the weight. However, with ratio data the results are more dependent on the size and configuration of the areas than when you use the total number in each area. That's because you're introducing a factor related to how an area was delineated—either the areal extent (for density) or the total population contained in the area (for percentage).

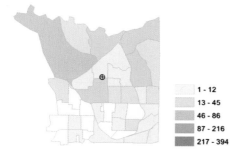

	1 - 12
	13 - 45
	46 - 86
	87 - 216
	217 - 394

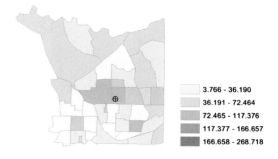

	3.766 - 36.190
	36.191 - 72.464
	72.465 - 117.376
	117.377 - 166.657
	166.658 - 268.718

TRACT	AREA (SQ MI)	HISPANIC	HISPANIC / SQ MI
0089	4.36	394	90.3
0090	1.28	86	67.1
0092	1.36	139	101.8
0091	1.27	122	96.4
0005	1.27	31	24.5
0001	1.25	74	59.1

The center of Hispanic population using population per block group (left map) and using density (population per square mile for each block group)

HOW THE MEAN CENTER IS CALCULATED

The mean center is the location represented by the mean x-coordinate value and the mean y-coordinate value for all the features in the study area. It's useful for tracking changes in the distribution or for comparing the distributions of different types of features. The center is calculated by summing the x-coordinate values and dividing the total by the number of features, and then doing the same for the y-coordinate values. The resulting x,y coordinate pair is the location of the mean center.

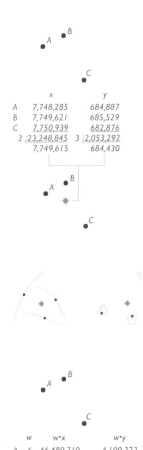

	x	y
A	7,748,285	684,887
B	7,749,621	685,529
C	7,750,939	682,876
3)23,248,845	3)2,053,292
	7,749,615	684,430

Sum the x-coordinate values (X) of the features
(i represents each individual observation, or feature)

The mean x-coordinate (the bar signifies mean value)

$$\bar{X} = \frac{\sum_i X_i}{n}$$

Divide the sum by the number of features (n)

$$\bar{Y} = \frac{\sum_i Y_i}{n}$$

The mean y-coordinate is calculated the same way

The locations of lines are represented by the x- and y-coordinates of the center point of the lines. Areas are represented by the x- and y-coordinates of their centroids.

The equation for calculating the weighted mean center is similar to the one for the mean center. Before being summed, each x-coordinate and each y-coordinate is multiplied by the weight for that feature. The result is divided by the sum of the weights, rather than the number of features in the set.

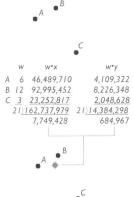

	w	w*x	w*y
A	6	46,489,710	4,109,322
B	12	92,995,452	8,226,348
C	3	23,252,817	2,048,628
	21)162,737,979	21)14,384,298
	7,749,428	684,967

The weighted mean x-coordinate (the w subscript stands for "weighted")

Multiply each x-coordinate (X) by that feature's weight (w), then sum the weighted coordinate values

$$\bar{X}_w = \frac{\sum_i (w_i X_i)}{\sum_i w_i}$$

Divide the summed coordinate values by the sum of the weights

$$\bar{Y}_w = \frac{\sum_i (w_i Y_i)}{\sum_i w_i}$$

The weighted mean y-coordinate is calculated the same way

The 3D mean center is calculated in the same way, with the addition of the mean z-value:

The mean Z value

$$\bar{Z} = \frac{\sum_i Z_i}{n}$$

Sum the Z values and divide by the number of features

The 3D center, then, is the mean center as it would be on the surface along with the mean z-value, the average distance above or below the surface.

As with the mean center, you can also calculate the 3D weighted mean center. To calculate the weighted mean center of a set of earthquakes, you'd use the depth as the z-value and the magnitude as the weight.

The mean center—whether weighted or not, in 2D or 3D—is the average of the coordinates of all the features in the dataset. The calculation doesn't directly consider the spatial relationships between the features, such as the distance between them (although the weighted center does take into account the relative value of the attribute used as the weight value).

HOW THE MEDIAN CENTER AND CENTRAL FEATURE ARE CALCULATED

The median center and central feature calculations both rely on the distance between features in the dataset. Because of this, these statistics are useful for finding the center when there is travel between it and the features.

Calculating the median center

There is no single equation for calculating the exact median center. The software approximates the center by iteratively calculating the mean center, summing the distances from it to each feature, offsetting the center slightly and summing the distances again. It eventually homes in on the location that has the lowest sum, which becomes the median center.

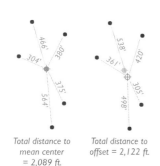

Total distance to mean center = 2,089 ft.

Total distance to offset = 2,122 ft.

Since the calculation attempts to minimize the total distance from all features, the center will gravitate toward the areas with the most features.

To calculate the weighted center of minimum distance, the GIS multiplies the distance between the interim center and each feature by the weight value for that feature. For a point with a weight of 6, the distance would be multiplied by 6 before being summed with all the other distances. The resulting distance is equal to having six points at that location (six trips to the center generated). A point on the opposite side of the study area with a weight of 3 would have its distance multiplied by 3. The weighted center would be pulled in the direction of the point with the greater weight.

Total weighted distance to center = 19,422 ft.

Calculating the central feature

Calculating the central feature is more straightforward than calculating the median center. The GIS totals the distance from each feature to every other feature. The feature with the lowest total distance to all other features is the central feature.

To calculate the weighted central feature, the GIS calculates the distance between each feature and each other feature, one pair at a time, and multiplies it by the weight value. These weighted distances are then summed—the feature having the lowest total weighted distance is the central feature.

FACTORS INFLUENCING THE LOCATION OF THE CENTER

If you're analyzing discrete features, the mean center or median center might not be at, or even near, any feature locations. This is especially true if the features are widely dispersed or if there are two or more localized clusters in the distribution. In the latter case, the center will fall somewhere between the clusters.

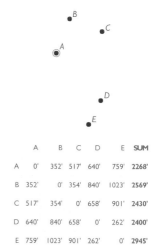

	A	B	C	D	E	SUM
A	0'	352'	517'	640'	759'	2268'
B	352'	0'	354'	840'	1023'	2569'
C	517'	354'	0'	658'	901'	2430'
D	640'	840'	658'	0'	262'	2400'
E	759'	1023'	901'	262'	0'	2945'

Mean center of businesses. In the map on the left, the businesses are widely dispersed and the center is not near any features. In the map on the right, the businesses cluster at opposite sides of the study area, and the center falls between the clusters.

One or more outliers can skew the mean center or median center. This is especially true if the dataset contains few features—the more features there are, the less influence any single feature will have on the sum of coordinate values. Outliers can also skew the weighted center—a feature with a very high value will pull the center toward it.

NAME	TYPE	EMPLOYEES
Classic Signs & Lettering	Services	1
Norman Shearer Construction	Construction	3
Optima Precision Inc	Wholesale	6
Optima Farm	Agriculture	1
Northern Medical	Services	860
W R Gamble Engineer	Services	1
Wankers Corner Saloon & Cafe	Retail	13
Starr Dental Ceramics	Services	1

In the map on the left (drawn using proportional symbols) and in the table, it's apparent that one business has a much larger number of employees than the surrounding businesses. The weighted mean center gravitates toward that business (right map).

An outlier may be a feature that is located incorrectly—especially if it was geocoded from a street address. Make sure the feature is in the correct location by comparing it to the source data (paper maps or files). If you have selected a subset of features (such as burglaries from a set of crimes), double-check that the feature is coded correctly.

It could be a feature with the wrong attribute value. The value may have been entered incorrectly and will need to be corrected by editing the layer's attribute table and entering the correct value. If the value is unknown, delete the feature from the dataset.

The feature might be an errant point. You may need to delete the feature from the analysis, or redefine the study area to exclude it.

A feature that doesn't fall in any of the above categories is likely a valid feature that should be included in the analysis.

For the weighted center, features having an attribute value of zero or having no value will not be included in the sum of coordinate values. Be sure the value of zero is valid. A count of zero for low-weight births may mean that there were none in a county, but it could also mean that no data was collected. If no data was collected, the calculated center will not be accurate. Update the dataset with the correct values, if available, or else exclude that feature from the analysis. Do the same for features having no value.

If you're calculating the weighted mean for contiguous areas, and there are many small areas located near each other, the center will be pulled in that direction—even if the weight values aren't high—because there are more features in that area.

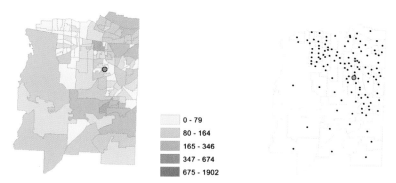

	0 - 79
	80 - 164
	165 - 346
	347 - 674
	675 - 1902

Mean center of population age 65 or older. The map on the right shows the block group centroids, which are used to represent the feature locations. Since the weighted center is calculated using the feature locations as well as the weights, it is pulled toward where there are more features, as well as toward features with high weight values.

Usually, multiple events at a single location are stored as individual features in the GIS database. This is particularly true of events that are geocoded to street addresses, as when several calls to 911 are made from the same house, often the case when there is an ongoing medical problem. Similarly, apartment buildings are usually geocoded as a single street address and hence a single location in the GIS. So calls to 911 from different apartments in a building would all be geocoded to the same location.

INCIDENT #	DATE	LOCATION
110561126	04/08/98	836 WEBSTER ST
110631531	04/15/98	827 POST ST
110641014	04/16/98	827 POST ST
110750741	04/27/98	840 E CITRUS AV
112502043	10/19/98	821 OXFORD DR
110011332	02/12/98	831 COLLEGE AV
110172301	02/28/98	840 E CITRUS AV
111771612	08/07/98	830 W BROCKTON AV
109971242	02/08/98	831 N SIXTH ST
112091833	09/08/98	907 W OLIVE AV
112890024	11/27/98	840 E CITRUS AV
112891131	11/27/98	840 E CITRUS AV

Incidents of domestic violence, with multiple incidents at some addresses

If you want the location of the center to be influenced by the locations of all events, having multiple events at a single location is not a problem. The center is automatically weighted toward the locations with the most events since each event is counted as a single feature. It's the same as counting the number of events at the location and using the count as a weight. You'd use this approach if, for example, you're calculating the mean center of incidents of domestic violence in order to scout possible locations for a counseling center. You'd include all calls from each address to weight the center toward the locations with the most calls.

1
5
10

Incidents of domestic violence symbolized to show the number of incidents at each address during one year, with the mean center of the incidents

If, however, you're concerned only with the locations where events occurred—and not with the number of events at each location—you'd want to calculate the center using each unique location rather than the number of events at each. For example, if you're holding a one-time workshop on domestic violence and you want to pick a location central to all households that reported at least one incident of domestic violence, you'd calculate the center using each household, rather than each call. You don't care how many calls were made from each household. In this case, you need to make sure there is only one record in the database for each address.

The mean center of locations experiencing at least one incident of domestic violence during one year

LOCATION	COUNT
836 WEBSTER ST	1
827 POST ST	2
840 E CITRUS AV	4
821 OXFORD DR	1
831 COLLEGE AV	1
830 W BROCKTON AV	1
831 N SIXTH ST	1
907 W OLIVE AV	1

Measuring the compactness of a distribution provides a single value representing the dispersion of features around the center. The value is a distance, so the compactness can be represented on a map by drawing a circle with the radius equal to the value.

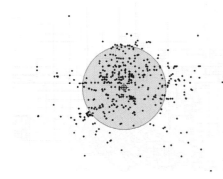

The dispersion of assaults in a city during one year, represented by a circle around the mean center

To calculate the compactness of a distribution, the GIS measures the average distance the features vary from the mean center. The measure is called the standard distance deviation, or simply standard distance.

You can use the standard distance values to compare two or more distributions. A crime analyst, for example, could compare the standard distance of assaults and auto thefts. If the distribution of crimes in a particular area is compact, stationing a single car near the center of the area might suffice. If the distribution is dispersed, having several police cars patrol the area might allow officers to respond more quickly.

Assaults (shown in red) are more concentrated than auto thefts (shown in blue), as shown by the relative size of the circles.

Comparing the same type of feature over different time periods—daytime and nighttime burglaries—would show if burglaries are more dispersed by day than by night.

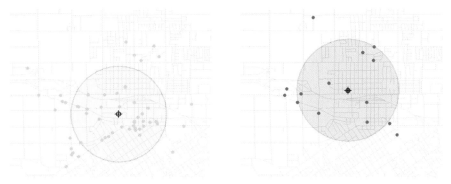

Daytime burglaries (left map) tend to be slightly more concentrated than nighttime burglaries in this city. Daytime and nighttime burglaries also tend to occur in different areas.

You can also compare the standard distance for distributions associated with stationary features. For example, you could measure the dispersion of emergency calls over several months for each fire station in a region and compare the standard distances to see which stations respond over a wider area.

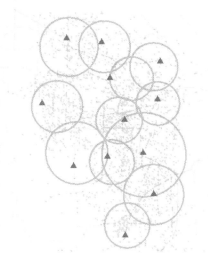

The standard distances of emergency calls that each fire station responded to, along with the locations of the stations (triangles) and calls (blue dots). The map shows the size of the response areas, where areas overlap, and where stations are relative to their areas.

CALCULATING THE STANDARD DISTANCE BY LOCATION OR BY ATTRIBUTE

You can calculate the standard distance for feature locations alone or influenced by attribute values associated with the features. The latter is termed the weighted standard distance. A city planner analyzing employment patterns could calculate the standard distance for a group of businesses, to see how dispersed or concentrated the businesses themselves are. Or, the planner could calculate the standard distance for the businesses weighted by the number of employees at each business. If the employees are more concentrated than the businesses, it would confirm that the largest employers are near the center of businesses.

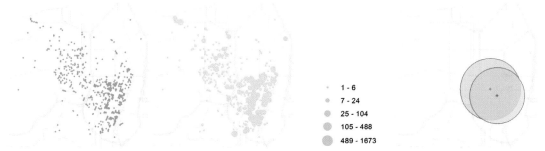

•	1 - 6
●	7 - 24
●	25 - 104
●	105 - 488
●	489 - 1673

Locations of financial businesses (left map) and financial businesses symbolized by the number of employees (center). The map on the right compares the standard distance for the business locations (3,034 feet) and the standard distance weighted by number of employees (2,679 feet). The distribution of employees is more compact than the distribution of businesses.

For contiguous areas you'd only calculate a weighted standard distance. A social scientist studying integration patterns could compare the standard distance for census tracts weighted by the population of different ethnic groups, such as African Americans and American Indians. That would show which group is more dispersed around its population center.

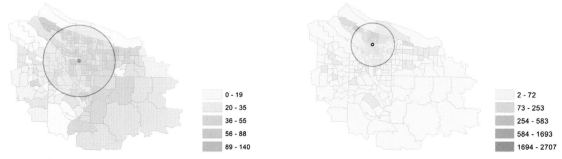

0 - 19		2 - 72		
20 - 35		73 - 253		
36 - 55		254 - 583		
56 - 88		584 - 1693		
89 - 140		1694 - 2707		

The weighted standard distance for the American Indian population by census tract is 35,685 feet (left map) in this region; for the African American population, the standard distance in the region is 21,621 feet.

	x	y	x-x̄	y-ȳ
A	7,748,285	684,887	-1330	457
B	7,749,621	685,529	6	1099
C	7,750,939	682,876	1324	-1554

	$(x-\bar{x})^2$	$(y-\bar{y})^2$
A	1768900	208849
B	36	1207801
C	1752976	2414916
	3)3521912	3)3831566
	1173971 +	1277189 = 2451159

$SD = \sqrt{2451159} = 1565.6$

WHAT STANDARD DISTANCE MEASURES

The standard distance statistic measures the extent to which the distances between the mean center and the features vary from the average distance.

Calculating the standard distance

The GIS first calculates the average difference in distance between the points and the mean center of the distribution (for lines and areas, the centroid of each feature is used). To do this, it subtracts the value of the mean x-coordinate from the x-coordinate value for each point and squares the difference (so the result will be a positive number). It then does the same for the y-coordinates. Once all the differences from the mean have been calculated, it sums them and divides by the number of points in the set.

Next, it sums the two resulting values, and takes the square root of the result to return the values to the original distance units. The resulting value is the standard distance.

Calculate the difference between each feature's x-coordinate and the mean x-coordinate (X-X̄), square the result, then sum the squared differences; do the same for the y-coordinates....

The Standard Distance

$$SD = \sqrt{\frac{\sum_i (X_i - \overline{X})^2}{n} + \frac{\sum_i (Y_i - \overline{Y})^2}{n}}$$

....divide each sum by the number of features (n), add the two, and take the square root

This is similar to calculating the standard deviation for a set of data values, which is the average amount the values differ from the mean. The difference here is, since you're dealing with geographic locations, there are two variables: the x-coordinate and the y-coordinate. By adding the distance for both and taking the square root, the GIS calculates a single distance value.

Calculating the weighted standard distance

The GIS calculates the weighted standard distance similarly to how it calculates the weighted mean center. It takes the squared difference in coordinate values between each point and the weighted mean center and multiplies it by the weight, sums the weighted differences, then divides the summed values by the sum of the weights.

The difference between each feature's coordinate and the mean coordinate value is multiplied by the feature's weight (w) and then summed....

The weighted Standard Distance

$$wSD = \sqrt{\frac{\sum_i w_i (X_i - \overline{X})^2}{\sum_i w_i} + \frac{\sum_i w_i (Y_i - \overline{Y})^2}{\sum_i w_i}}$$

....then the sum of the weighted differences is divided by the sum of the weights

X and Y are the x- and y-coordinates of the weighted mean center.

INTERPRETING THE RESULTS OF THE STANDARD DISTANCE CALCULATION

The result of the standard distance calculation is a distance in whatever units the coordinates of the geographic features are in (such as feet or meters). The greater the standard distance value, the more the distances vary from the average, and the more widely dispersed the features around the center.

You can graphically represent the standard distance by using the calculated distance as the radius to draw a circle around the mean center.

The standard distance for sightings of white-tailed deer in this study area is 26,973 meters.

When you draw the standard distance circle, you'll see that some points will be inside the circle and some outside. Features inside the circle vary less than the standard distance from the mean. Feature outside vary more.

The standard distance provides a better measure of compactness for features distributed regularly around the mean, rather than clustered at opposite sides of the study area.

The map on the left shows the standard distance for sightings of badgers—the sightings are fairly evenly distributed around the mean center. The map on the right shows the standard distance for sightings of bobcats in the same area—the sightings are clustered at opposite sides of the study area. (In some cases there are several sightings at the same location.)

Standard distance works best when there is no strong directional trend. If there is a directional trend, use the standard deviational ellipse (discussed in the next section), which measures both compactness and orientation.

The standard distance circle and standard deviational ellipse for sightings of sandhill cranes in one study area

Measuring orientation and direction lets you abstract the spatial trends in a distribution of features. By calculating the orientation of a group of burglaries, for example, a crime analyst could see that the pattern of crimes follows the trend of major streets. That might provide clues about the location and the type of establishments that attract burglars (it might be easier to go undetected on major streets as well as get away quickly).

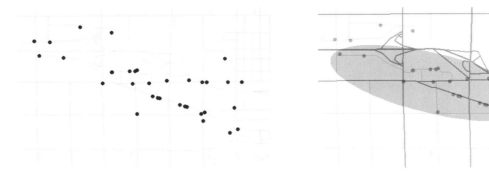

On the left, burglaries; on the right, the orientation of burglaries (indicated by the ellipse), with major streets

Measuring the orientation of points and areas is different than for line features. For points and areas, you use a measure similar to the standard distance circle that calculates the variance separately for the x-coordinates and y-coordinates. The result can be displayed as an ellipse showing the orientation of the distribution. A wildlife biologist studying how bobcats move between preferred habitat areas in a region could calculate the orientation of the areas to see if their orientation coincides with natural features such as valleys, rivers, or ridgelines. If so, it would provide clues to the travel routes the bobcats use.

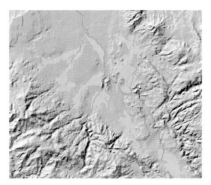 

Bobcat habitat areas on the left; the orientation of habitat areas on the right

To see the orientation of line features, you calculate a line showing the average orientation of all the line features. For example, a geologist could calculate the orientation of fault lines in an area to compare the trend of the lines to geologic patterns under the surface.

Orientation (blue line) of fault lines (color coded by type)

MEASURING THE ORIENTATION OF POINTS AND AREAS

Measuring the trend for points or areas is usually done by calculating the standard distance separately in the x and y directions. These two measures define the axes of an ellipse encompassing the distribution of features. The ellipse is referred to as the standard deviational ellipse since the method calculates the standard deviation of the x-coordinates and y-coordinates from the mean center. The ellipse allows you to see if the distribution of features is elongated and hence has a particular orientation.

While you can get a sense of the orientation by drawing the features on a map, calculating the standard deviational ellipse makes the trend clear. It also gives you confidence in your analysis since the result is based on a statistical calculation and not just on a visual interpretation of the map.

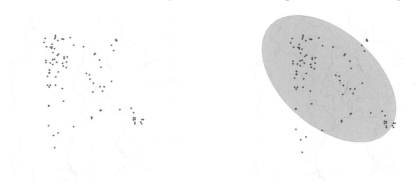

Moose sightings (left map) and the standard deviational ellipse for the sightings

Calculating the ellipse gives a more accurate picture than using the standard distance circle. Consider two distributions that create identical ellipses that have the same center, except one is oriented north–south and the other is oriented east–west. Clearly they don't coincide very well, but the standard distance circles for the distributions would perfectly overlap.

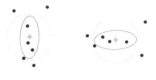

The standard deviational ellipse is useful for comparing the distributions of categories of features. For example, an economic development planner might want to see whether different types of businesses—such as financial and manufacturing—have similar trends, defining an economic corridor for the city.

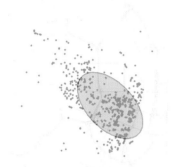 

The standard deviational ellipses for financial businesses (left map) and manufacturing businesses in a region

It is also useful for comparing one type of feature at different times. For example, a wildlife biologist studying moose behavior would map the ellipses for moose observations during various months and compare them to see the corridors moose use in different seasons.

 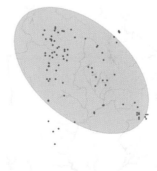

Ellipses for sightings of moose in a study area for March (left map) and September

Similarly, an epidemiologist could map the ellipses week by week for cases of a disease in a village to summarize the spread of the disease.

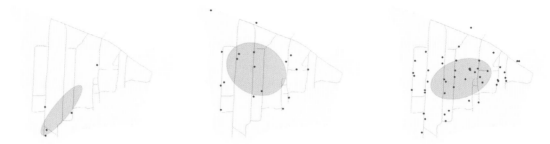

Ellipses for cases of dengue fever in a village for week two, week three, and week four of an outbreak

As with the standard distance, you can calculate the standard deviational ellipse using either the locations of the features or using the locations influenced by an attribute value associated with the features. The latter is termed a weighted standard deviational ellipse.

Standard deviational ellipse for locations of financial businesses (left map), and the ellipse for businesses weighted by the number of employees at each (shown using proportional symbols). The ellipse indicates larger financial businesses are concentrated in a district trending north to south.

What the standard deviational ellipse statistic measures

The standard deviational ellipse measures the standard deviation of the features from the mean center separately for the x-coordinates and the y-coordinates.

Calculate the difference between each feature's x-coordinate and the mean x-coordinate (X-X), square the result, then sum the squared differences....

The Standard Distance for the x-axis

$$SD_X = \sqrt{\frac{\sum_i (X_i - \overline{X})^2}{n}}$$

....divide the sum by the number of features (n), and take the square root

$$SD_Y = \sqrt{\frac{\sum_i (Y_i - \overline{Y})^2}{n}}$$

The Standard Distance for the y-axis is calculated the same way

The equation for the weighted standard deviational ellipse is similar, except the squared difference from the mean for each coordinate is multiplied by the attribute value before being summed.

The lengths of the axes are calculated in the east–west (x-axis) and north–south (y-axis) directions, in distance units (such as feet or meters). Since the standard deviation is measured in each direction from the mean center, the total length of each axis is twice its standard deviation.

To determine the orientation of the ellipse, the GIS employs a trigonometric function. It calculates an angle of rotation from 0° (due north) for the y-axis so that the sum of the squares of the distance between the features and the axes is minimized. It then rotates each axis by this angle. Essentially, it tries to find the best fit of both axes among the features to minimize the distance of the features to the axes.

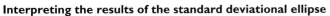

$SD_{major} = 182$ ft.
$SD_{minor} = 80$ ft.

Interpreting the results of the standard deviational ellipse

The GIS displays the values of the coordinates of the mean center, the standard distance of each axis, and the angle of rotation of the ellipse. You may also have the option of drawing the axes and the outline of the ellipse.

The standard deviational ellipse for these businesses has an x-axis standard distance of 2,630 feet, a y-axis standard distance of 1,502 feet, and an angle of rotation of 313° from north.

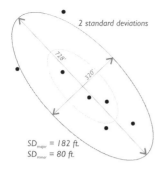

2 standard deviations

SD_{major} = 182 ft.
SD_{minor} = 80 ft.

As with the standard distance circle, since the standard deviation represents the average distance features vary from the mean center, features at a greater distance than this average will fall outside the ellipse.

You can also calculate and draw an ellipse using two, or more, standard deviations. A biologist studying a particular plant species could calculate the ellipses for known observations using several standard deviations and concentrate a search for other occurrences of the species within that area.

An ellipse calculated using one standard deviation will show where the features are concentrated. An ellipse calculated using two or more standard deviations will show you where most of the features occur.

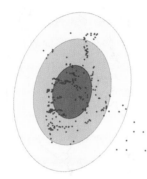

The ellipses calculated using one, two, and three standard deviations, for a distribution of moose sightings in a study area.

The orientation or size of the ellipse can be skewed by a few outlying features, and thus not provide an accurate picture of the distribution. This is especially true if the dataset contains few features and there are several outliers near each other.

The ellipses for a group of burglaries with an outlier (left), and without the outlier (right)

MEASURING THE DIRECTION AND ORIENTATION OF LINE FEATURES

The trend of a set of line features is measured by calculating the average angle of the lines. The statistic used to calculate the trend, though known as the directional mean, is used to measure either direction or orientation.

Many linear features point in a direction—they have a beginning and an end. Such lines often represent the paths of objects that move, such as hurricanes. Features like fault lines, with no start and end point, are said to have an orientation, but no direction. You can calculate either the mean direction or mean orientation of a set of lines.

Mean orientation for a set of fault lines

You can calculate the directional mean for either discrete features, such as storm tracks, or for features connected in a network, such as streams or rail lines.

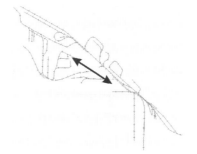

Mean direction of cyclone storm tracks (left), and mean orientation of rail lines through a region

Measuring the trend is useful for comparing two or more sets of lines. For example, a wildlife biologist studying the movement of elk and moose in a stream valley could calculate the directional trend of migration routes for the two species.

Mean direction of migration paths for elk (left) and moose. While the paths indicate differences in movement, the overall direction for both species is similar.

Alternatively, a climatologist studying tropical storms in the Pacific Ocean could calculate the directional trend of storm tracks during one season to compare the general trend of storms in different parts of the ocean.

Mean direction of storm tracks over one season in the western Pacific Ocean (red line) and eastern Pacific (blue line)

Calculating the trend is also useful for comparing features for different time periods. For example, an ornithologist could calculate the trend for falcon migration month by month. The directional mean summarizes the flight paths of several individuals and smoothes out daily movements. That makes it easy to see during which month the birds travel farthest.

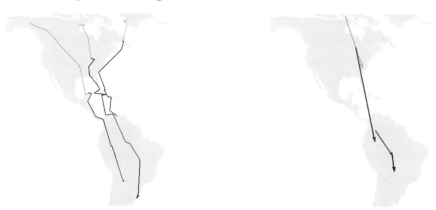

Paths of three tundra peregrine falcons migrating from North to South America over four months (left map), and the mean direction of the paths for all three birds, for each month

Calculating the directional mean lets you compare the trend in a set of lines to other features to look for possible relationships. For example, you could calculate the directional mean for a stream network and compare it to the standard deviational ellipse for a species' habitat areas. If the habitat areas' orientation corresponds to the direction of streams, you could then see if the streams influence the locations of the habitat. While you could get a sense of this simply by mapping the streams and habitat areas, calculating the direction and orientation quantifies the trends and makes the relationships more apparent.

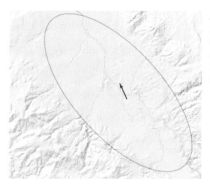

Mean direction of streams (blue arrow) with the standard deviational ellipse showing the orientation of bobcat habitat areas

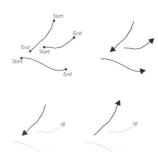

Measuring direction or orientation

In a GIS, every line is assigned a start point and an end point, and thus has a direction. The direction is set when the line feature is created, by digitizing or by importing a list of coordinates. You can see the direction of each line by displaying it with an arrowhead symbol.

If you're calculating the mean direction, you need to make sure the directions of the lines are correct, and fix them if they aren't. Most GIS software lets you "flip" lines to reverse the start and end points.

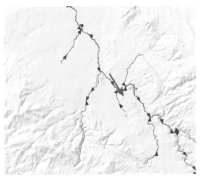 

In the map on the left, some stream segments correctly point downstream, while others point upstream. When the mean direction is calculated, the arrow points upstream. In the map on the right, the segments pointing upstream have been flipped, so that all segments point downstream, and the mean direction arrow now also points downstream.

If you're calculating the mean orientation, the direction of the lines is ignored.

In general, the mean direction is calculated for features that move from a starting point to an end point, such as storm tracks, while mean orientation is calculated for stationary features, such as fault lines.

Mean direction for storm tracks (left map) and mean orientation for earthquake faults (right)

There may be situations where you'll want to calculate the mean orientation of lines that represent movement. While a wildlife biologist interested in where elk start and end during their seasonal migration would calculate the mean direction of the paths the elk take during each season, a biologist interested in what makes a good route would calculate the mean orientation using the elk paths in both directions.

Mean direction for seasonal elk migration paths through a valley (left) and mean orientation of elk migration paths in a study area over the course of several years (right)

What the directional mean statistic measures

The directional mean statistic is the angle of a line that represents the mean direction (or orientation) of all the lines in the dataset. Unlike other calculations of the mean, the angles can't be summed and divided by the number of features to get the mean angle, since the values are on a circular scale (degrees) rather than a linear one. For example, if you had three lines in the dataset, one pointing north–northeast (20° from due north), one pointing slightly more to the north (15°), and one pointing slightly west of north (355° from due north), the sum of the angles would be 390, and the mean angle would be 130°—a bearing of southeast, and clearly not the mean direction of the three lines.

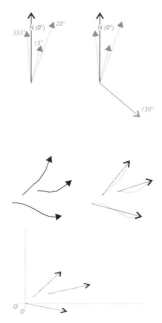

To calculate the angle, the features are all placed in the same "graph space," with the origin at 0,0.

The lines are considered to have two points, start and end—vertices within the line are ignored. Only direction matters, so for the calculation, each line is considered to be one unit long. (It doesn't matter what the units are since the lines are in graph space.) At this point the lines are called vectors or unit vectors.

The mean direction is calculated by adding the vectors together and measuring the angle of the line that connects the origin to the end point of the last vector. Doing this manually—as described by geographers Jay Lee and David Wong—lets you see how the directional mean is derived. First, you place a vector's start point at the origin of the graph, then lay the rest of the vectors end to end in any order, maintaining the original direction

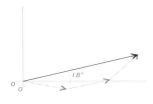

of each. The order in which the lines are appended doesn't matter—
the resulting end point will be the same. Draw a line (the resultant vector) from the origin to the end point of the last vector. Finally, you'd use a
protractor to get the angle of the resultant vector—this angle is the directional mean. (In graph space, the x-axis is assigned 0°, and the angles are
calculated counterclockwise from this direction. This is different than geographic space, where 0° is due north.)

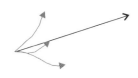

If you placed the origin of each line at the same location and kept the
direction the same, you'd expect the vector representing the directional
mean to fall somewhere in the middle of all the lines. In order to be able
to create the resultant vector and measure its angle, though, the lines have
to be turned into unit vectors and appended.

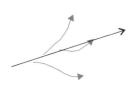

The GIS does the same thing, using trigonometry to calculate the various
angles. The GIS calculates the angle of each line from the coordinates of
the beginning and end points. Once it has all the angles it calculates the
sine and cosine of each and sums these. It divides the sum of sines by the
sum of cosines, and calculates the arc tangent of the result, which is the
angle of the resultant vector—the directional mean.

The angle of the resultant vector (i.e., the directional mean) · *Sum the sines of the angles of the line features (from the horizontal), and divide by....*

$$\theta_R = \arctan \frac{\sum_i \sin\theta_i}{\sum_i \cos\theta_i}$$

....the sum of the cosines of the angles of the features; then take the arctangent of the result

The arctan function returns a value up to 180°. To get the correct angle
in a 360° range, the software applies a correction based on whether the
numerator and denominator of the equation are positive or negative values. Essentially, it shifts the resultant vector into the correct quadrant
(northwest, southwest, or southeast) if necessary.

When you calculate the mean orientation, the GIS makes any lines pointing in opposite directions point the same direction. Two earthquake
faults with the same southwest–northeast orientation would have different angles if they were pointing in different directions (because of the way
they were digitized). One would have an angle of 45° and the other an
angle of 225°. Calculating the mean of these angles would result in a vector with an angle of 135°, a southeast direction, and not correctly indicating the trend of the fault lines.

Interpreting the results of the directional mean calculation

The result is reported in degrees counterclockwise from due east in geographic space—a legacy of trigonometry, in which angles are calculated counterclockwise from the positive x-axis in graph space. Some GIS software, such as ArcGIS, gives you this angle as well as the more traditional compass direction, calculated clockwise from due north. In addition to reporting the angle, ArcGIS also creates a line feature using the directional mean value and locates it at the mean center of the lines, so you can graphically see the directional mean.

The mean direction of these elk migration paths is 341°, measured counterclockwise from the x-axis (due east). It's 109° clockwise from north.

Factors influencing the directional mean statistic

Since the angles of the lines are added together, lines pointing in different directions and having widely varying angles can cancel each other out. The directional mean will point in a direction between the lines and will not provide a valid measure.

Bighorn sheep migration paths in a study area. Since the paths point in various directions, the directional mean does not provide useful information.

The direction of the lines is important, since that's how the angle is calculated (if the line is pointing in one direction, the angle will be the reverse of the angle if it were pointing in the other direction). You need to make sure the direction of each line is correct.

Lines with a similar orientation may point in opposite directions. This is not a problem if you're measuring mean orientation. If you're measuring mean direction, however, you may need to separate the lines based on the direction they're pointing, and calculate the mean direction for each set.

Mean direction for a set of elk migration paths. Since the paths point both east and west, the mean direction ends up pointing northwest. The map on the right shows the mean orientation for the same set of paths—the orientation arrow matches the east–west trend of the paths.

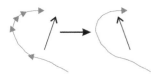

Sometimes what appears to be a single line is actually stored in the GIS as several individual lines, connected at nodes. When calculating the directional mean, the GIS counts each line separately. If you don't want to calculate the directional mean using the individual lines, but rather the entire line from end point to end point, you'll need to merge the individual lines to create a single line.

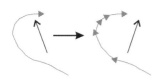

Conversely, if you have a single line with twists and turns, the GIS will only use the direction from the start point to the end point when calculating the mean direction. If you want to calculate the mean direction using the changes in direction within the line, you'll need to split it into separate lines.

Measuring the variability in direction or orientation

Knowing the extent to which the line features vary from the mean tells you how confident you can be that the directional mean reflects the actual trend of the lines. The circular variance, derived from the directional mean and calculated using the length of the resultant vector, tells you the extent to which features all point in the same direction. If, for example, all the lines pointed east, the directional mean would also point east. But if half the lines pointed northeast and half pointed southeast, the directional mean would still point east, even though this distribution of linear features is very different from the first. However, the circular variance for the two sets of lines would be much different.

The circular variance of the bighorn sheep migration paths in the study area on the left is 0.97; the variance of the paths in the study area on the right is 0.08—the lower the number, the less variation in direction between the lines.

In order to calculate the circular variance, you need to first calculate the length of the resultant vector.

Sum the sines of the angles of the line features (from the horizontal) and square the sum; then do the same for the cosines....

The length of the resultant vector

$$OR = \sqrt{\left(\sum_i \sin\theta_i\right)^2 + \left(\sum_i \cos\theta_i\right)^2}$$

....then add the two and take the square root

On the graph, you can see that the sum of sines and the sum of cosines represent the two sides of the triangle formed by the resultant vector and the intersections of its end point with the x and y axes. This equation for calculating the resultant vector is simply the Pythagorean theorem for calculating the hypotenuse of a triangle (the resultant vector): $c = \sqrt{(a^2 + b^2)}$.

A dataset having more lines could have a greater length for the resultant vector than a dataset having fewer lines even if the variability in direction is the same. The resultant vector for a dataset with five lines all pointing in the same direction would be longer than for a dataset with three lines all pointing in the same direction—but the variability is the same in both cases (in both sets, all lines point in the same direction). To account for the difference in the number of lines, the length of the resultant vector is divided by the number of vectors. The result is then subtracted from 1 so that a small value represents low variability.

The circular variance

$$S = 1 - \frac{OR}{n}$$

OR Divide the length of the resultant vector (OR) by the number of features (n)....

....then subtract the result from 1

The circular variance is reported as a number between 0 and 1. If all lines pointed in the same direction, the length of the resultant vector would be equal to the number of features (since each vector is one unit long). In that case OR/n would equal 1, and the circular variance would be 0. On the other hand, if the vectors all pointed in opposite directions, the length of the resultant vector would be 0 (it would just be a point at the origin). In that extreme case, $0/n = 0$, and $1-0 = 1$; the circular variance would be 1.

The circular variance for the faults in the study area on the left is 0.43; for the faults in the study area on the right, it's 0.03.

So the closer the circular variance to 0, the more the lines point in the same direction, and the stronger the trend. The closer the circular variance to 1, the more variability there is in the line directions, and the weaker the trend.

Burt, James E., and Gerald M. Barber. *Elementary Statistics for Geographers.* Guilford Press, 1996. The chapter on descriptive statistics covers mean center, directional mean, and other measures. Discussions are concise, with an emphasis on the theory and math behind the statistics.

Earickson, Robert J., and John M. Harlin. *Geographic Measurement and Quantitative Analysis.* Macmillan, 1994. The "Measurement and sampling" and "Descriptive statistics" chapters provide straightforward discussions of the basics of measuring distributions, including mean center and standard distance.

Ebdon, David. *Statistics in Geography.* Blackwell, 1985. The "Spatial Statistics" chapter includes a discussion of mean and median center, standard distance, and standard deviational ellipse, illustrated with excellent diagrams. An introductory chapter covers the basics of statistical concepts.

Lee, Jay, and David W. S. Wong. *Statistical Analysis with ArcView GIS.* Wiley, 2001. Includes sections on calculating the mean and median center for point distributions, dispersion of point distributions, and directional statistics for distributions of line features. Lee and Wong show calculations for each statistic using sample data.

Levine, Ned. *CrimeStat: A Spatial Statistics Program for the Analysis of Crime Incident Locations (v 2.0).* Ned Levine & Associates and the National Institute of Justice, 2002. The CrimeStat documentation covers the concepts and theory behind the statistics. Includes examples of the application of the statistics to crime analysis.

Wong, David W. S. "Geostatistics as Measures of Spatial Segregation." *Urban Geography* 20, no. 7 (1999): 635–47. In this article, Wong proposes a method for statistically comparing two or more distributions using the standard deviational ellipse, and provides a case study of the method applied to analyzing the segregation of ethnic groups in a city.

Testing statistical significance

The statistics in chapters 3, 4, and 5 allow you to identify a probable pattern or relationship. Before making a decision that involves committing resources or before doing further analysis, you'll want to know with some degree of certainty that your conclusions about the pattern or relationship are correct. Significance tests give you a probability that what a statistic is telling you is true. Probability is a measure of chance, and underlying all statistical tests are calculations that assess the role of chance on the outcome of your analysis.

The assaults look clustered, but you don't know if the clustering is due to chance, unless you test the statistical significance of the pattern.

Statisticians distinguish between analyzing all possible observations within a study area (the population) and analyzing a sample. In the social sciences, since you can't talk to every person in a large area, predictions or assumptions are made about a group of people by collecting information from a few. In the natural sciences, samples are collected when you can't observe every individual. For example, when studying the distribution of plants in a region, it would be difficult to identify and note the location of every single plant of a particular species, so you'd set up a number of transects across the region and sample the plants along each transect. You'd then make estimates about the distribution of all the plants in the region from your sample.

The goal is to have the results of analysis for the sample match as closely as possible the results you would get had you been able to analyze the entire population. When a sample is a good representation of the population as a whole, you can expect measures for the sample to be near those of the population. Any difference you find is called sampling error. How well the sample matches the population depends on the characteristics of the sample, its size, and how it was collected.

The sample you collect is only one of many possible samples. If you obtained a different sample, you might very well draw a different conclusion about the population. Most samples—if they're large enough—would produce a good estimate of the characteristics of the population. Now and then you would, by chance, draw a sample that produced results very different from the true results for your study area; but the probabilities of drawing a highly unusual sample set would be very slim.

STATISTICAL SIGNIFICANCE, PROBABILITY, AND SAMPLING

The smaller the sample, the more likely that one or a few extreme values included by chance will skew the results. With a larger sample, any extreme values that happen to be included in that particular sample will have a smaller effect on the distribution.

To predict with some degree of certainty, inferential statistics require that each observation has an equal and independent chance for inclusion in a sample—the selection of one observation should in no way influence the selection of the next observation. Samples that meet this requirement are said to be random. If a large enough random sample is collected, you can calculate the probability that the sample you've obtained is a good estimate of the population as a whole.

CONFIDENCE LEVELS AND NULL HYPOTHESES

Statistical theory provides methods to measure levels of confidence associated with observations you make. The observations or beliefs are stated as a hypothesis. It's only human to favor any patterns or relationships you see or expect to see. So, to maintain impartiality, you set out to prove the opposite—the so-called null hypothesis. Your initial hypothesis is called the alternative. Significance tests help you decide whether you should or should not reject the null hypothesis.

In order to decide whether to reject the null hypothesis, you first decide the risk you are willing to accept for being wrong. This degree of risk, often referred to as the confidence level (or significance level), is expressed as a probability ranging from 0.0 to 1.0. The desired level of confidence is compared with an observed level of confidence to decide whether to reject or not to reject the null hypothesis. The observed level of confidence, known as the p-value, is calculated using your sample data. So the characteristics of your sample (its size, whether it's a random sample) affect the observed level of confidence.

The most common confidence levels for statistical tests are 0.10, 0.05, and 0.01. These values indicate the risk you are willing to accept for erroneously rejecting the null hypothesis. If the study could be repeated 100 times, each time with a different sample, probability indicates that, at the 0.05 confidence level, 95 out of 100 studies would yield accurate results. In other words, 5 out of 100 would likely yield erroneous results due to sampling error.

Each statistical tool has an appropriate significance test (sometimes more than one). The test provides a statistic that represents the p-value. And, the desired confidence level has a corresponding critical value (which depends on the specific test). If the value of the test statistic exceeds the critical value, you reject the null hypothesis. The results of your analysis are said to be statistically significant at the specified confidence level.

Rather than saying they accept the null hypothesis, statisticians say that they "don't reject" it. That's because you're never absolutely sure (it's all about probabilities), so you're not in a position to absolutely accept anything. Further analysis or new information may show that you should reject the null hypothesis. Similarly, rejecting the null hypothesis doesn't prove that the alternative hypothesis is true—it only indicates that your sample data does not support the null hypothesis.

Most spatial statistics tools calculate a test statistic at the same time they calculate the initial statistic and report both. Many of the tools calculate a Z-score, a reference measure for the standard normal distribution (with mean of zero and standard deviation of 1).

The critical value for the Z-score at a confidence level of 0.05 is 1.96. If the Z-score is within the range −1.96 to 1.96, the null hypothesis cannot be rejected. If it falls outside this range, you can reject the null hypothesis.

The critical values of −1.96 and 1.96 are standard deviations from the mean. Ninety-five percent of the area underneath the standard normal curve falls between plus and minus 1.96 standard deviations from the mean. The other 5% of the area is termed the rejection region. If the Z-score falls within the rejection region, there's only a 5% chance that you'd be wrong to reject the null hypothesis.

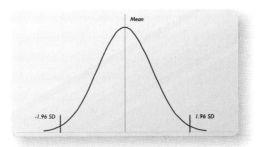

The critical values for Z at a confidence level of 0.05 are plus and minus 1.96 standard deviations from the mean—95% of the area under the curve (representing 95% of all value) falls within these limits.

At a confidence level of 0.01, 99% of the area under the curve falls between −2.58 and 2.58 standard deviations of the mean.

Confidence level	Z-score critical values
0.01	±2.58
0.05	±1.96
0.10	±1.65
0.20	±1.28

These tests were developed within classic (nonspatial) statistics. Spatial data, however, contradicts some of the assumptions of inferential statistics, by its very nature. You need to be aware of the limitations of the tests, and the assumptions the tests make, and question the results. The result of the significance test—as is the statistic you're testing and, indeed, as is a map—is just one piece of information to aid in making a decision.

The tests are based on estimating the probability that the variation in your sample reflects the variation in the population as a whole. And they assume you're testing a random sample. In GIS analysis, though, it's often the case that the data has already been collected and put into a database. You may not know if the data was randomly sampled or how large the sample is compared to the population; you may not even know if you are dealing with a sample or with the whole population.

Even if you assume you are analyzing a sample, spatial data often violates one of the main assumptions of statistics—independence of observations in a sample. In a hypothetical random spatial distribution, every feature or observation would have an equal probability of occurring at any given position, and the position of any given feature or observation would have no influence on any other feature or observation in the dataset. But, the selling price for one particular house, for example, does in fact influence nearby housing values.

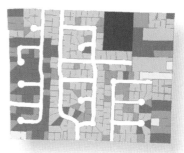

High-value properties (dark blue) are likely to be found together.

Similarly, the observations are often not random—commercial burglaries occur only where there are businesses, and businesses tend to cluster, so the outcome is somewhat predetermined. Also, spatial data is rarely evenly distributed across a region—for example, rainfall may increase as you move from west to east.

Rainfall in this region increases from west to east.

Finally, it's not uncommon in the GIS setting to find yourself working with very large datasets. Obtaining statistical significance in these cases will not be difficult if you're analyzing several thousand features or more.

Increases in computing power are allowing researchers to explore methods that rely on computer-generated distributions derived from millions of permutations over a dataset, such as Monte Carlo simulation and Bootstrapping. These methods, which may eventually replace theoretical distribution models and their accompanying mathematical equations, avoid the issues associated with statistical tests based on sampling (such as assumptions of normality or spatial independence). Nonetheless, the Z-score test and other tests based on sampling are still commonly used to assess statistical significance for spatial data.

The null hypothesis for spatial pattern analysis is that the features are evenly distributed across the study area. However, in many cases, it's difficult to envision a situation in which the null hypothesis could or would ever be true. Suppose, for example, you have urban employment data and want to measure whether or not the locations of employment opportunities are clustered—it would be difficult to imagine an urban environment in which employment opportunities didn't cluster.

Locations of employers

While performing a statistical test to show that your employment data exhibits statistically significant clustering would not tell you anything you didn't already know, it would confirm your observations and understanding of what's happening. That could be important if decisions made using the analysis have legal or economic ramifications. Probably even more useful, though, would be to see if any strong clustering that is also statistically significant increases or decreases over time. That would tell you if employment is becoming more or less concentrated, and would provide insights into structural changes taking place in the city.

If you're analyzing the distribution of feature values, such as the median house value by census tract in a county, you have to make some assumptions about how the values were sampled (since the significance tests assume you're working with sample data). Two common sampling assumptions are randomization and normalization.

Census tracts for a county, color coded by median house value

Randomization sampling assumes that the observed spatial pattern of your data represents one of many possible spatial arrangements—the number of features having a particular value is always going to be the same (based on the observed number of each), but the arrangement can change. Suppose you have the number of cases per census tract of a disease. Some tracts have several cases of the disease; many don't have any. Your null hypothesis would be that the disease strikes randomly. If you could take the case count values for your study area and scatter them on a map of the census tracts, making sure every census tract got a value, you'd create a random pattern. The randomization null hypothesis postulates that if you could perform this operation infinite times, most of the time you would produce a pattern that was not markedly different from the observed pattern. If your significance test indicates that you should reject the null hypothesis at the specified confidence level, then you know that the observed arrangement of values would significantly differ from this randomly produced pattern.

Normalization sampling, in contrast, assumes that the number of cases associated with any particular census tract could be derived from an infinitely large, normally distributed population of values (through some random sampling process). Rather than scattering the observed values on the map of census tracts, you'd pick values from this hypothetical normal distribution and scatter those values on the map to create the random pattern.

With the randomization null hypothesis, you know up front all possible values; with the normalization null hypothesis, you don't know all possible values—you assume that the values are a sample from a larger population. That's hard to conceptualize, especially when you're analyzing spatially continuous data or data summarized by contiguous areas. If your study area is a county, for example, it would be hard to make the case that the set of values for all census tracts in the county is a sample.

The normalization null hypothesis not only assumes that your data is a sample, but also that the sample was obtained randomly and that the population from which the sample was obtained has a normal distribution of values. Every time you make an assumption about the data or the sample, you're potentially introducing error into the test. Randomization

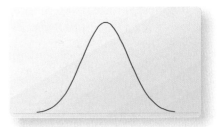

Frequency curve for a normal distribution

makes fewer assumptions than normalization, so it's safer to use, unless you know for sure your data matches the assumptions of normalization.

The Z-score (or other test result) will be calculated differently depending on the assumption, and hence the resulting value will be different. Since you compare the Z-score value to a critical value at a given significance level, the assumption you use has a bearing on whether the result is significant or not at that level.

Some software calculates and presents the results using both assumptions, in which case you'll need to decide which assumption best matches your situation, and use that result. In other cases, you may need to specify the assumption to use for the test statistic.

You don't truly know if the null hypothesis is true or false—you decide to either reject it or not, with some level of confidence. You can try to reduce the risk that you'll draw the wrong conclusion from the significance test by minimizing the likelihood you'll make an error in rejecting the null hypothesis.

If you reject the null hypothesis and it actually is false, or—again based on your test—you don't reject it and it actually is true, then your conclusion is correct. But if you reject the null hypothesis and it's actually true, or you don't reject it and it's actually false, you're making an error. Statisticians call the first one a Type I error, and the other a Type II.

The risk is usually less with a Type I error. Suppose, for example, you're trying to identify potential environmental factors that may be contributing to a particular type of cancer. Your null hypothesis states that these cancer incidents are distributed randomly throughout your study area. If you commit a Type I error (you reject the null hypothesis, falsely concluding that cancer incidents are clustered), you will likely move to a more specific level of analysis, such as examining relationships among the cancer clusters and particular environmental factors, or seeing if you can reproduce your findings in a second study area. Hopefully, upon further scrutiny, your error will be uncovered. On the other hand, if you commit a Type II error (you fail to reject the null hypothesis, falsely concluding that this particular type of cancer strikes randomly), you may abandon your research prematurely. The bigger risk in this case lies with committing a Type II error.

Because with spatial statistics the null hypothesis is already established, the best you can do to favor a Type I error is specify a less stringent confidence level. If you use a confidence level of 0.1 instead of 0.05, you're more likely to say there's significant clustering (that is, reject the null hypothesis). If there actually isn't clustering (the null hypothesis is true), at worst you've only committed a Type I error. If you set a higher confidence level (say 0.01), you're less likely to say there's significant clustering (that is, you're more likely to accept the null hypothesis). If there is in fact clustering, you've committed a Type II error.

While statistical significance gives you some confidence that you have found a pattern or a probable relationship, it's not the final answer. There may be larger questions that factor into any decisions you make using the results of your analysis, such as whether your findings have important implications or even whether you're asking the right question in the first place. Regardless of the results of your test, you'll want to consider these larger issues when assessing the results of your analysis.

References

Bailey, Trevor C., and Anthony C. Gatrell. *Interactive Spatial Data Analysis*. Longman, 1995.

Burt, James E., and Gerald M. Barber. *Elementary Statistics for Geographers*. Guilford Press, 1996.

Earickson, Robert J., and John M. Harlin. *Geographic Measurement and Quantitative Analysis*. Macmillan, 1994.

Ebdon, David. *Statistics in Geography*. Blackwell, 1985.

Goodchild, Michael F. *Spatial Autocorrelation*. Catmog 47, Geo Books, 1986.

Griffith, Daniel A., and Carl G. Amrhein. *Statistical Analysis for Geographers*. Prentice-Hall, 1991.

Hamilton, Lawrence C. *Regression with Graphics: A Second Course in Applied Statistics*. Brooks/Cole, 1992.

Lee, Jay, and David W. S. Wong. *Statistical Analysis with ArcView GIS*. Wiley, 2001.

3 Identifying patterns

Identifying the pattern created by a set of features allows you to gain a better understanding of the distribution of features, monitor conditions, compare different features, or track changes.

In this chapter:

- Why identify geographic patterns?
- Using statistics to identify patterns
- Measuring the pattern of feature locations
- Measuring the spatial pattern of feature values

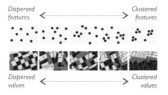

Dispersed features ← → Clustered features

Dispersed values ← → Clustered values

Any distribution of features or attribute values within a defined area will create a pattern. Geographic patterns range from completely clustered at one extreme to completely dispersed at the other. A pattern that falls at a point between these extremes is said to be random.

Knowing if there's a pattern in your data is useful if you need to gain a better understanding of a geographic phenomenon, monitor conditions on the ground, compare patterns, or track changes.

BETTER UNDERSTANDING OF GEOGRAPHIC PHENOMENA

If a biologist studying wildlife behavior found that family groups in an area are dispersed, it could mean that the species can live in a wide range of habitats. If the groups are clustered, it could mean that the species has very specific habitat requirements and lives only where that habitat exists.

Western grebe habitat is dispersed across Wyoming (left), while red-necked grebe habitat is clustered (shown using 5-km raster cells).

MONITORING CONDITIONS

A forestry agency responsible for monitoring timber harvest practices in an area could measure the pattern of clear cuts to ensure there is sufficient contiguous forest habitat remaining. The agency could specify an allowable level of clustering of clear cuts and then regularly measure the pattern to make sure that level isn't exceeded.

The level of clustering of cut areas (dark green) within a forest can be measured to monitor compliance.

COMPARING DIFFERENT SETS OF FEATURES

A crime analyst could compare the patterns of different crimes, such as burglaries, assaults, and auto thefts. If the crimes create a clustered pattern, the analyst could identify hot spots and commanders could assign patrols to those areas. If the crimes are dispersed, something like a neighborhood watch program could be implemented.

The patterns of burglaries (blue), assaults (red), and auto thefts (orange) can be compared in an area to tailor tactics to the type of crime.

TRACKING CHANGE

Public health officials studying the spread of a disease could measure the pattern of where new cases of the disease occur. If the pattern becomes more clustered over time, it could indicate the disease is spreading less rapidly.

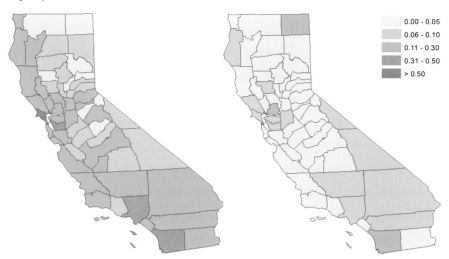

Counties color coded by new cases of AIDS per 1,000 people in 1994 (left) and 2003. In 1994, counties having many new cases were dispersed across the state. There were fewer new cases by 2003 and counties with new cases were clustered, for the most part.

One way of identifying patterns in your geographic data is to display the features or values on a map. Another way is to use statistics to measure the extent to which features or values are clustered, dispersed, or random. With that measure you can easily compare the patterns for different sets of features or compare patterns over time.

When you use statistics to measure patterns, you compare the actual distribution of features (often referred to as the observed distribution) to a hypothetical random distribution of the same number of features over the same area. The GIS calculates the statistic for the observed distribution as well as what the statistic would be for a random distribution. The extent to which the observed distribution deviates from the random distribution is the extent to which the pattern is more clustered or more dispersed than the random distribution.

The process of measuring a pattern by comparing an observed distribution to a random distribution is based on the scientific method, wherein you state the null hypothesis and then set out to determine whether to reject it, or not, with some level of confidence. Because of your knowledge of the features and how they behave, you probably already have an idea that the features you're analyzing are either clustered or dispersed, or not. You can use statistics to confirm your idea and to measure how strong the pattern is—the degree of clustering or dispersion.

The results of the analysis can be tested to calculate the probability that a pattern isn't simply due to chance. Calculating this probability is important if you need to have a high level of confidence in any decision you make—for example, if there are public safety or legal implications. (See "Testing statistical significance.")

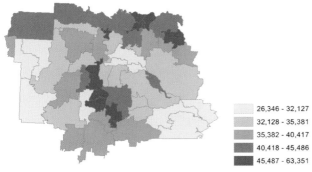

	26,346 - 32,127
	32,128 - 35,381
	35,382 - 40,417
	40,418 - 45,486
	45,487 - 63,351

Using statistics, you can be 95% sure the clusters of ZIP Codes having high median income did not occur by chance.

Using statistics to measure patterns is more accurate than identifying patterns by looking at a map. On a thematic map, the classification method you use, the number of classes, and the class ranges can all affect whether there appears to be a pattern or not. Since the statistical tools use the underlying data value of each feature, identifying a pattern does not depend on how the values are classified and displayed.

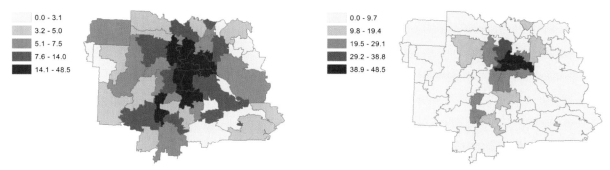

| 0.0 - 3.1 |
| 3.2 - 5.0 |
| 5.1 - 7.5 |
| 7.6 - 14.0 |
| 14.1 - 48.5 |

| 0.0 - 9.7 |
| 9.8 - 19.4 |
| 19.5 - 29.1 |
| 29.2 - 38.8 |
| 38.9 - 48.5 |

Percent Hispanic population by ZIP Code using quantile classification (left) and equal interval classification. The population appears more dispersed when the quantile classification is used.

If point features are close together on the map you may not see that there are many features grouped together—they may appear as a single feature. Similarly, multiple events occurring at a single location, such as several emergency calls from a single address, will also appear as a single feature. So, it's hard to discern whether the features form a clustered or dispersed pattern. Using graduated symbols can help reveal the patterns, but this solution is not ideal since the symbols may themselves overlap. Using statistics to measure the distribution of features—and the pattern they form—avoids this problem since each individual feature is counted in the calculation.

Locations of burglaries (left), and the same data mapped by number of burglaries at each address, using graduated symbols (right)

GEOGRAPHIC SCALE AND MEASURING PATTERNS

The pattern created by a set of features may change depending on the scale. For example, the features may be dispersed when you're conducting analysis on a small study area, but clustered when you're analyzing the same set of features within a larger study area.

Great Plains toad habitat in Wyoming is clustered when the entire state is the study area, but is dispersed if the study area is confined to the county in the northeast corner, as shown in these raster datasets.

If you have a large study area with a lot of local variation, measuring the pattern may be meaningless. For example, analyzing the pattern of clear cuts for North America would not yield useful information, since clear cuts would likely match the pattern of certain forest types. On the other hand, analyzing the pattern of clear cuts in western British Columbia, where the forest is fairly homogeneous, would give you some insight into timber practices in that region.

THE RELATIONSHIP BETWEEN PATTERN AND LOCAL VARIATION

Statisticians distinguish between measures to identify and quantify the pattern created by the features in the study area and measures that identify variation across the study area. The former—termed global statistics—focus on whether or not the features form a pattern across the study area, and on what type of pattern it is. The latter—termed local statistics—focus on individual features and their relationship to nearby features. Local statistics are discussed in chapter 4, "Identifying clusters."

Local statistics are useful if you need to identify the locations of clusters of features. Using a local statistic can help find hot spots when a global statistic indicates that there is a clustered pattern.

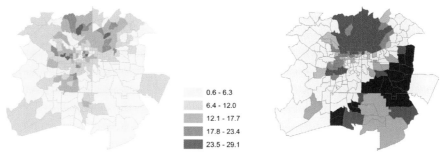

	0.6 - 6.3
	6.4 - 12.0
	12.1 - 17.7
	17.8 - 23.4
	23.5 - 29.1

Global statistics indicate whether there is a pattern (in this case, for the percentage of seniors in each census tract); local statistics identify clusters of features with similarly high values (orange) or low ones (blue).

Sometimes the global and local statistics don't agree. For example, a global statistic could indicate no significant clustering, whereas at a local scale you can identify clusters. Using both global and local statistics together provides a better sense of what's actually occurring in the study area.

MEASURING THE PATTERN OF FEATURE LOCATIONS VERSUS FEATURE VALUES

The statistical tools for measuring patterns globally fall into two categories. One group of tools lets you measure the pattern formed by discrete features, such as points, lines, or noncontiguous areas.

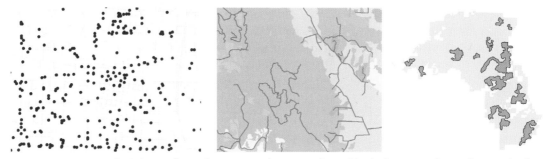

Statistics can be used to measure the pattern formed by the locations of point features (such as emergency calls), line features (such as logging roads), or discrete areas (such as clear cuts in a forest).

The other group includes tools that let you measure the spatial pattern formed by attribute values associated with the features. These tools are mainly used for analyzing contiguous areas (such as census tracts).

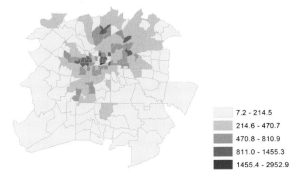

	7.2 - 214.5
	214.6 - 470.7
	470.8 - 810.9
	811.0 - 1455.3
	1455.4 - 2952.9

Population over 65 years of age, per square mile

The pattern formed by the distribution of points, lines, or discrete areas can be measured by overlaying areas of equal size, calculating the average distance between features, or counting the number of features within a defined distance.

Overlaying areas of equal size

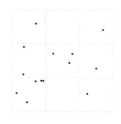

In this method, termed quadrat analysis, the GIS overlays areas of equal size on the study area and counts the number of features in each. It also calculates what you'd expect the counts to be for a hypothetical random distribution. If fewer areas than expected contain most of the features and more areas than expected contain few or no features, the features form a clustered pattern.

Use quadrat analysis if there is no direct interaction between the features—for example, when analyzing the distribution of a plant species. This method also works well for multiple events at a single location, such as several emergency calls to 911 from the same address, since it relies on the number of features in each area and not the distance between them. It's also appropriate when you have data that is already summarized by equal-size areas, so you don't have the x,y, coordinate location of each feature (as might be the case if you have raster data with counts of plants for each cell).

Calculating the average distance between features

This method, referred to as the nearest neighbor index, measures the distance between each feature and its nearest neighbor, and then calculates the average. Distributions that have a smaller average distance between features than a random distribution would have are considered clustered.

Use the nearest neighbor index if there is direct interaction between the features—for example, if you're analyzing the pattern of incidents of a contagious disease where infected people have had contact with others. This method is also useful when you're analyzing data that is distributed in a line, such as a set of plant or wildlife observations collected along a transect, since the statistic is calculated using the distance between features—the other methods base their calculations on the areal extent of the study area. There are, however, additional issues involved when analyzing line features (see "Using statistics with geographic data").

Counting the number of features within defined distances

In this method, you specify a distance interval, and the GIS calculates the average number of neighboring features within the distance of each feature. Increasing distances at the specified interval show at what distance the concentration of features is greatest. The method is known as the K-function since the distance ranges are represented by the variable K. It is also sometimes referred to as Ripley's K-statistic after statistician Brian Ripley. If the average number of features found at a distance is greater than the average concentration of features throughout the study area, the distribution is considered clustered at that distance. You can plot the statistical value at each distance to see if there is clustering at various distances.

Use the *K*-function if you're interested in how the pattern changes at different scales of analysis. For example, you would use the *K*-function to find out at what distance burglaries are most clustered.

OVERLAYING AREAS OF EQUAL SIZE

In this method, the GIS lays a set of equal-size areas, termed quadrats, over the study area and counts the number of features in each quadrat. This method originated in field research for analyzing the distribution of point features such as plants or archaeological artifacts. Traditionally, squares are used for the shape of the quadrats, in a regular grid (hence the name "quadrat"). That's because when quadrats were laid out in the field, a regular grid was the easiest to create with stakes and string. However, other shapes such as circles or hexagons are sometimes used. The quadrats can be contiguous (as in a grid) or distributed randomly across the study area.

The quadrat size will affect whether any patterns are identified. The goal is to choose a size that is large enough to capture any pattern but not so large it obscures the pattern. If quadrats are small, many of them may have few points or none. A large quadrat size may result in quadrats with about the same number of points in each.

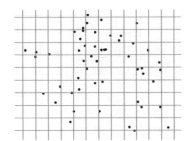
Quadrat size (per side) = 2,000 ft.; many quadrats contain no points.

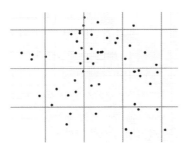
Quadrat size = 6,000 ft.; quadrats have close to the same number of points.

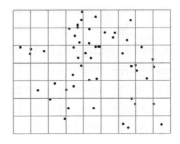
Quadrat size = 3,000 ft.; quadrats are large enough so many contain at least one point, but small enough to provide a range in the number of points per quadrat.

If possible, you should use the characteristics of the features you're analyzing to determine the quadrat size. For example, creosote bushes in the southwestern United States often grow in a dispersed pattern—the less rainfall there is in an area, the farther apart plants will be. Using a quadrat size less than the average distance between plants in a given study area would result in many quadrats with no features in them.

If you have few features and the features are spread out, you may need to use larger quadrats. If you have many features, and they are concentrated in some areas, you may need to use smaller quadrats.

Some researchers suggest that the quadrat size equal twice the size of the mean area per feature, which is simply the area of the study area divided by the number of features. In most GIS software, you specify the quadrat size by giving the length of one side of the square, so you would take the square root of the result.

The length of a side of a quadrat

Divide the extent of the study area (A) by the number of features (n), multiply by 2...

$$l = \sqrt{2 \cdot \frac{A}{n}}$$

....and take the square root

$$l = \sqrt{2 \cdot \frac{139392000}{200}} = 1180$$

For a study area that is 2 miles in the east–west direction and 2.5 miles in the north–south direction, containing 200 points, the mean area per point is 696,960 square feet, the area of each quad is 1,393,920 square feet, and the length of a side is 1,180 feet.

You can also run the analysis using quadrats of different sizes to see how much the results differ. This will also tell you if the pattern changes at different scales. For example, features may be dispersed locally but clustered on a regional scale (revealed using larger quadrats).

What quadrat analysis measures

Because the method counts the number of features per unit area, it measures the density of the features. It doesn't take into account the proximity of features to each other, or their arrangement in relation to each other, as do the nearest neighbor index and the *K*-function.

Calculating the quadrat analysis statistic

The GIS counts the number of features in each quadrat and creates a frequency table. The table lists the number of quadrats containing no features, the number containing one feature, two features, and so on, all the way up to the quadrat containing the most features.

The GIS then creates the frequency table for the expected distribution, usually based on a Poisson distribution (see "Understanding data distributions").

# points	# quads
0	34
1	11
2	7
3	5
4	4
5	1
6	1

To do this, the GIS calculates the probability that a given number of features will occur in a quadrat, first by finding the average number of features per quadrat, referenced as the Greek letter lambda.

The average number of features per quadrat (lambda)

Divide the number of features (n) by the number of quadrats (k)

$$\lambda = \frac{n}{k}$$

The GIS then uses lambda to determine the probability of a particular number of features occurring in any given quadrat.

The probability (p) of "x" number of features occurring in a quadrat

e is Euler's constant, with a value of 2.7183; it's raised to the power of negative lambda....

....and multiplied by lambda raised to the x power (if you're calculating the probability of 3 features occurring in a quadrat, lambda is cubed)....

$$p(x) = \frac{e^{-\lambda} \lambda^x}{x!}$$

....then the result is divided by x factorial (3! = 3 × 2 × 1 = 6); x! is shorthand for the permutations of the numbers up to and including x (the numbers 3, 2 and 1 can be ordered six different ways)

The GIS uses the equation to calculate the probability for 0 features occurring, 1 feature occurring, 2 features, and so on, and lists these probabilities in the frequency table. It then multiplies the probability for each class by the total number of features to get the number of quadrats expected to contain that number of features.

# points	p	expected # quads
0	.3905	24.60
1	.3672	23.13
2	.1726	10.88
3	.0541	3.41
4	.0127	0.80
5	.0024	0.15
6	.0004	0.02

By comparing the two frequency tables, you can see whether the features create a pattern. If the table for the observed distribution has more quadrats containing many features than does the table for the random distribution, then the features create a clustered pattern.

It's hard to judge the extent to which the frequency tables are similar or different just by looking at them. Plus, the tables may resemble each other just by chance—that is, in any given random distribution a clustered or dispersed pattern can occur just by chance (see "Testing statistical significance"). So, there are several statistical tests you can use to find out how much the frequency tables (and thus the patterns) differ.

# points	observed # quads	expected # quads
0	34	24.60
1	11	23.13
2	7	10.88
3	5	3.41
4	4	0.80
5	1	0.15
6	1	0.02

Testing the results of the quadrat analysis

Two commonly used tests are the Kolmogorov-Smirnov test, which compares the difference in frequencies between the tables, and the Chi-square test, which compares the cumulative difference in frequencies.

Kolmogorov-Smirnov

The Kolmogorov-Smirnov test, named after two Russian mathematicians who developed similar equations in the 1930s, calculates the proportion of quads in each class for each line in the frequency table (the number of quads in the class divided by the total number of quads). It then creates a running cumulative total of proportions from the top of the table to the bottom—the total for all classes equals 1 (100% of the quads).

The software does this for both the expected and random distributions. It then compares the running cumulative totals at each class level and selects the largest absolute difference, referenced as *D*.

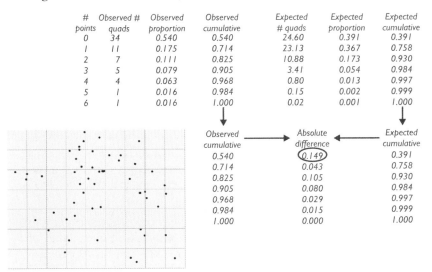

# points	Observed # quads	Observed proportion	Observed cumulative	Expected # quads	Expected proportion	Expected cumulative
0	34	0.540	0.540	24.60	0.391	0.391
1	11	0.175	0.714	23.13	0.367	0.758
2	7	0.111	0.825	10.88	0.173	0.930
3	5	0.079	0.905	3.41	0.054	0.984
4	4	0.063	0.968	0.80	0.013	0.997
5	1	0.016	0.984	0.15	0.002	0.999
6	1	0.016	1.000	0.02	0.001	1.000

Observed cumulative	Absolute difference	Expected cumulative
0.540	0.149	0.391
0.714	0.043	0.758
0.825	0.105	0.930
0.905	0.080	0.984
0.968	0.029	0.997
0.984	0.015	0.999
1.000	0.000	1.000

Calculating D for 63 quads covering 67 features

It compares this value to the critical value for a confidence level you specify. The software calculates the critical value based on the confidence level and the number of quads.

The critical value for a given confidence level is calculated by
dividing the constant (K) by the square root of the number of quads (m)

$$\text{critical value} = \frac{K}{\sqrt{m}}$$

K is a mathematical constant, the value of which depends on the confidence level you're using. The value of K is found in a standard table for the Kolmogorov-Smirnov test. For a confidence level of 0.20 (80%), K is 1.07. So, for example, if you have 63 quads, the critical value at a confidence level of 0.20 would be

$$\text{critical value} = \frac{1.07}{\sqrt{63}} = \frac{1.07}{7.90} = .135$$

If the calculated difference in proportions between the observed and random distribution (D) is greater than the critical value, the difference can be considered statistically significant.

Chi-square

Also based on the frequency table, the Chi-square test (X^2 test) is often used to find out whether two sets of frequencies (or two distributions) are significantly different, for example, when comparing an observed distribution to a random distribution. If the differences between the two sets are small, the differences may simply be due to chance, and you can assume the distributions are not significantly different.

For each frequency level or class, the software calculates the difference between the observed number of quads and the expected number. It then squares the difference and divides by the expected number of quads. Finally, it sums the values to arrive at the test statistic.

At each frequency level (k), the expected number of quads (f_E)
is subtracted from the observed number of quads (f_O). The difference
is squared and divided by the expected number....

$$X^2 = \sum_1^k \frac{\left(f_O - f_E\right)^2}{f_E}$$

....then the results are then summed for all frequency levels.

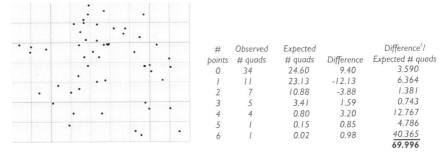

# points	Observed # quads	Expected # quads	Difference	Difference^2/ Expected # quads
0	34	24.60	9.40	3.590
1	11	23.13	-12.13	6.364
2	7	10.88	-3.88	1.381
3	5	3.41	1.59	0.743
4	4	0.80	3.20	12.767
5	1	0.15	0.85	4.786
6	1	0.02	0.98	40.365
				69.996

X^2 *calculated for 67 features covered by 63 quads*

You then compare the test value with the critical value, obtained from a standard table, which is in turn based on an expected distribution (the X^2 distribution). In the X^2 distribution, the highest numbers of occurrences are low values, with fewer occurrences as the values increase. This type of distribution is appropriate to test against when occurrences with high values are relatively rare, as is the case with quadrat analysis, where there may be many quads with a few features, and only a few quads containing many features. If the X^2 value is greater than the critical value, the distributions are significantly different, and you can consider the observed distribution to be not random, but either clustered or dispersed.

For any given confidence level, there is a range of critical values that depend on the degrees of freedom for your data sample, and the test. Establishing degrees of freedom allows for differences in sample size—with a smaller sample you have to be more cautious in rejecting the null hypothesis. The smaller the sample size, the fewer the degrees of freedom and the more stringent the critical value. To get the degrees of freedom, you start with the sample size (or the number of frequency classes in the case of the Chi-square test) and subtract based on the assumptions of the particular test. In this example, two degrees of freedom are lost—one because once you know how many quads are in the first six classes, you can predict the last class (there is only one quad, and six points, left), and another because the mean for the expected distribution (that is, lambda) was calculated from the data.

The Chi-square test requires a minimum number of quadrats in each feature class. One rule of thumb is that no more than 20% of the frequency classes contain fewer than five quadrats. A common workaround is to group frequency classes to ensure the minimum is met. In the example above, three out of the seven classes (42%) contain fewer than five quadrats, so you'd have to group the last three classes. However, you lose one degree of freedom each time you eliminate a class (since you're reducing the sample size), so the test becomes less useful. If you use Chi-square with large samples, you can avoid this grouping and get more accurate results than if you use it with small samples. The Kolmogorov-Smirnov test can be used with small samples, since it doesn't require a minimum number of quadrats in each class.

Factors influencing the results of the quadrat analysis

Quadrat analysis works best with small, tight clusters, as is the case with earthquakes or crimes such as commercial burglaries.

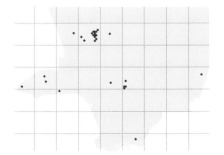

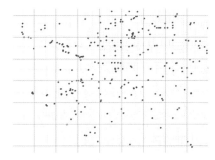

Earthquakes in Los Angeles County in March 1974

Commercial burglaries over one year

Since it doesn't measure the relationship or distance between features— merely whether they fall in the same quadrat—quadrat analysis may not recognize certain patterns. For example, a cluster of five points that happens to fall at the center of a quadrat would give the same measure as if the five points had been dispersed across the quadrat.

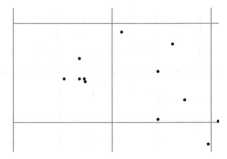

Quadrat analysis treats both quads as equal, since they both have the same number of points.

CALCULATING THE AVERAGE DISTANCE BETWEEN FEATURES

The nearest neighbor index is based on the research of ecologists Philip Clark and Frances Evans, who developed the method in the 1950s to quantify patterns in distributions of various plant species. In this method, the GIS finds the distance between each feature and its closest neighbor, then calculates the average (or mean) of these distances.

What the nearest neighbor index measures

The nearest neighbor index measures how similar the mean distance is to the expected mean distance for a hypothetical random distribution. The index is either the difference between the two or the ratio of the observed distance divided by the expected distance.

Calculating the observed mean distance

To get the distance from each feature to its nearest neighbor, the GIS essentially calculates the distance from each feature to all other features in the set, then finds the shortest distance, the nearest neighbor to the feature.

	A	B	C	D	E	F	G
A		988	2117	2494	3538	3858	4267
B	988		1725	3348	4308	4004	3601
C	2117	1725		2804	4034	2567	2309
D	2494	3348	2804		1196	2510	4897
E	3538	4308	4034	1196		3277	6034
F	3858	4004	2567	2510	3277		3433
G	4267	3601	2309	4897	6034	3433	

The software adds the distances between each pair of nearest neighbors, then divides by the number of features in the set to get the mean distance:

The mean distance for the observed distribution of features

Measure the distance to each feature's nearest neighbor (c_i), and sum the distances....

$$\overline{d}_o = \frac{\sum_i c_i}{n}$$

....then divide by the number of features

Calculating the expected mean distance

Using the equation above, the mean distance for a completely clustered distribution is 0 (all points are at the same location so the distance between each point and its nearest neighbor is 0).

For a completely dispersed distribution—as Clark and Evans demonstrated—the mean distance is the inverse of the square root of the number of points divided by the areal extent of the study area.

The expected mean distance for a completely dispersed distribution

$$\overline{d}_e = \frac{1}{\sqrt{n / A}}$$

Divide the number of features by the area of the study area, and take the square root...then divide the result into 1

The mean distance for a random distribution is halfway between the values for a completely clustered distribution and a completely dispersed distribution, 0 and 1, respectively. That is, the square root of the number of points divided by the area is then divided into 0.5 instead of 1.

The expected mean distance for a random distribution

$$\overline{d}_e = \frac{0.5}{\sqrt{n / A}}$$

Divide 0.5 by the square root of the number of features divided by the area

$$\overline{d}_e = \frac{0.5}{\sqrt{200 / 139\,392\,000}} = 416.7$$

Calculating the index

If the mean distance for the features you're analyzing is less than the mean distance for a random distribution, you can conclude that the observed distribution is more clustered than random. If the distance is greater, the distribution is more dispersed.

To calculate the index, the GIS subtracts the expected mean distance from the observed mean distance.

Subtract the mean distance for an expected random distribution from the mean distance for the observed distribution

The nearest neighbor index as the difference

$$d = \overline{d}_o - \overline{d}_e$$

- If the observed and expected means are equal, the difference is 0 and the observed distribution is random difference.

- If the expected mean is greater than the observed mean, the difference is less than 0 (a negative number) and the observed distribution is clustered.

- If the expected mean is less than the observed mean, the difference is greater than 0, and the observed distribution is dispersed.

An alternate way of creating the index is to calculate the ratio between the two mean distances.

The nearest neighbor index as a ratio

$$r = \frac{\overline{d}_o}{\overline{d}_e}$$

Divide the observed mean distance by the expected mean distance for a random distribution

- If the means are the same, the ratio is 1 and the observed distribution can be considered random.
- If the expected mean is greater than the observed mean, the ratio is less than 1 and the observed distribution is clustered. The closer to 0 (the value for a completely clustered pattern), the more clustered the pattern.
- If the expected mean is less than the observed mean, the difference is greater than 1 and the observed distribution is dispersed.

Difference	Ratio	Pattern
$d < 0$	$r < 1$	Clustered
0	1	Random
$d > 0$	$r > 1$	Dispersed

The nearest neighbor index considers the relationship between features, unlike quadrat analysis which simply considers whether or not a feature falls within a particular area. That better captures any pattern when you're analyzing features that interact, such as cases of a contagious disease.

Testing the results of the nearest neighbor index
The null hypothesis is that the features are randomly distributed. A spatially random distribution would result in a normal distribution of the nearest neighbor distances. To help you decide whether or not to reject the null hypothesis—and conclude that there is a pattern other than random—the GIS calculates a Z-score. The Z-score test divides the difference between the observed and expected values by the standard error.

The expected mean distance is subtracted from the observed mean distance....

The Z-score

$$Z = \frac{\overline{d}_o - \overline{d}_e}{SE}$$

....and the difference divided by the standard error

The standard error measures the distribution of mean distances around their average value. This is equivalent to the standard deviation of the mean distance for the full set of features divided by the square root of the number of features. This equation is used to calculate the standard error:

The standard error

0.26136 *(a mathematical constant) is divided by....*

$$SE = \frac{0.26136}{\sqrt{n^2 / A}}$$

....the square root of the number of features squared (n²) divided by the areal extent (A) of the study area

The value 0.26136 is a constant derived from the radius of a circle, the notion for the standard error being based on using a circle divided into equal sectors and finding the number of points, given a hypothetical random distribution, in any given sector.

Since the difference is positive if the observed mean distance is greater than the expected mean distance, a positive Z-score indicates a dispersed pattern; conversely, the difference is negative if the observed mean distance is less than the expected mean distance; hence a negative Z-score indicates a clustered pattern. At a confidence level of 95%, the Z-score would have to be greater than 1.96 or less than −1.96 to be statistically significant. See "Testing statistical significance" for more on the Z-score.

Using the nearest neighbor index to compare distributions
You can compare the indexes for two distributions to see if one set of features is more or less clustered than another set. For example, you might calculate the nearest neighbor index for assaults and auto thefts, and then compare them to see if one is more clustered than the other. If assaults are more clustered, police could assign a task force to look at causes.

With a nearest neighbor index ratio of 0.42, assaults in this area (left) are slightly more clustered than auto thefts (right), which have a ratio of 0.50.

You can use the same approach to compare the observed distribution to a control distribution. For example, a forestry agency monitoring timber harvesting could compare logged watersheds to watersheds containing harvestable timber (the control distribution) to see if the logged watersheds are more clustered than you would expect, given the distribution of timber. The logged watersheds may be clustered because the timber company is harvesting forest stands that are closer than they should be. But they may be clustered simply because the harvestable timber is also clustered. This is especially true in mountainous areas where there may be rock outcrops or other barren land between the forest patches.

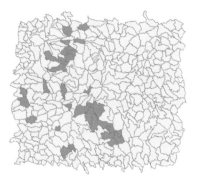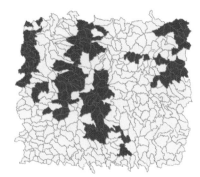

The logged watersheds (left) are not significantly more clustered than the watersheds containing productive forest (right).

You can also compare distributions from different study areas. However, unless the study areas are the same size (same areal extent), the results will not be valid.

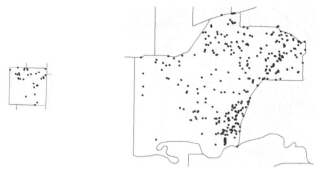

Emergency calls within two districts. The district on the left is 1.5 square miles, while the one on the right is 10.6. Comparing the nearest neighbor index for the two districts is misleading since the index depends on the areal extent of the study area.

Measuring the pattern at different relative distances

You can use the nearest neighbor index to find out if the pattern changes at different distances.

In the map of emergency calls on the left, the clusters are localized and the clusters themselves are close to randomly distributed. In the map on the right, the clusters form a dispersed pattern, so it might be effective to station response units at regular distances.

To see if the pattern changes, you calculate the index using the second-nearest neighbor, or the third-nearest, fourth-nearest, and so on. All the second-nearest neighbors constitute an order of neighbors. The different orders are referenced by the variable *k,* so this approach is sometimes called the *k*-order nearest neighbor index.

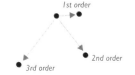

For the *k*-order index, the GIS calculates the observed mean distance in the same way as for the nearest neighbor, using the distance to the *k*-order neighbor. The expected mean distance for a random distribution is calculated using a variation on the equation presented earlier, with the factorial of the *k*-order:

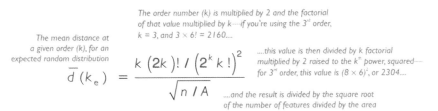

The order number (k) is multiplied by 2 and the factorial of that value multiplied by k—if you're using the 3rd order, k = 3, and 3 × 6! = 2160....

The mean distance at a given order (k), for an expected random distribution

$$\overline{d}(k_e) = \frac{k\,(2k)!\,/\,\left(2^k\,k!\right)^2}{\sqrt{n\,/\,A}}$$

....this value is then divided by k factorial multiplied by 2 raised to the kth power, squared—for 3rd order, this value is (8 × 6)², or 2304....

....and the result is divided by the square root of the number of features divided by the area

This is the full equation to calculate the expected mean distance for a random distribution—the simplified version shown earlier is how it looks when 1 (the first-order nearest neighbor) is substituted for *k*.

You can chart the orders and the index values as a line graph, orders along the x-axis and index values along the y-axis. If the slope of the curve increases steeply at first, and then levels out at a certain order, the points are locally clustered.

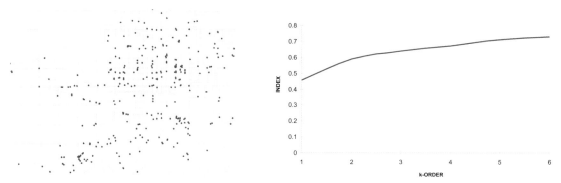

The chart shows that assaults are most clustered at the first-order neighbor (where the index is lowest) and decrease rapidly at the second order. Clustering then decreases gradually at higher orders.

Using a chart also makes it easy to compare the observed distribution to a random distribution. Since the index for the expected distribution is always 1 at each k-order (the expected mean distance is divided into itself), the curve for the expected distribution will appear as a straight line. The curve for the observed distribution will either be below this line (more clustered than random), above the line (more dispersed than random), or similar to the line (close to random). The curve might also start below the line and go above it (locally clustered and regionally dispersed), or start above the line and go below it (locally dispersed and regionally clustered).

You can also use the chart to compare the distributions of two or more subgroups of features (such as burglaries and auto thefts), or the same type of feature at two or more time periods. This will show whether one distribution is more or less clustered than the other at different relative distances.

Factors influencing the nearest neighbor index

For lines, usually either the midpoint on each line or a randomly chosen point is used to calculate the distance between features, depending on the software you're using. The results of the analysis could be misleading if there are situations where the end points are close but the midpoints aren't.

If what appear to be single continuous lines are actually several joined lines, the GIS may find the nearest neighbor to be the adjacent joined line. If that's not what you intend, you'll need to unsplit the lines.

For discrete areas, you can have the GIS use either the centroid of each area, or the locations where the boundaries are closest. If the areas are large, convoluted, long and narrow, or of different sizes, area boundaries may be quite close while the centroids are far apart. Using the centroids in these cases will likely show less clustering than actually exists. For convoluted areas, the centroid may even be outside the boundary. Using centroids works best when the areas are small, close to round or square, and close to the same size. (See "Using statistics with geographic data.")

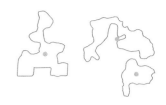

Features close to each other or at the same location can skew the mean distance to be smaller than it otherwise would be. Performing the analysis using *k*-order neighbors can present a more accurate measure of the distribution, since the second- or third-nearest neighbor will be at a greater distance.

INCIDENT	TYPE	ADDRESS
110351808	BURG, UNL ENT, COM DAY	1520 INDUSTRIAL PARK AV
110412204	BURG, FORCED, RES UNK	1547 WEBSTER ST
110261936	BURG, UNL ENT, RES DAY	1520 DWIGHT ST
110262125	BURG, FORCED, RES NITE	1547 WEBSTER ST
110421358	BURG, FORCED, RES DAY	1606 INDEPENDENCE AV
110382149	BURG, UNL ENT, COM NITE	1550 E HIGHLAND AV
110291050	BURG, UNL ENT, COM DAY	1520 INDUSTRIAL PARK AV
110201240	BURG, UNL ENT, COM DAY	1526 BARTON RD

Burglaries during March 1998. In some cases, more than one burglary occurred at the same address.

The extent of the study area can also affect the results, since it is used to calculate the expected mean distance. The software you're using may set the extent to the envelope created by the distribution of features or the full extent of the map view. You may have the option of specifying the area of the polygon that defines the study area, such as a county or census tract.

Area = 3.08 sq mi
Index = 0.35

Area = 1.46 sq mi
Index = 0.51

Area = 0.96 sq mi
Index = 0.63

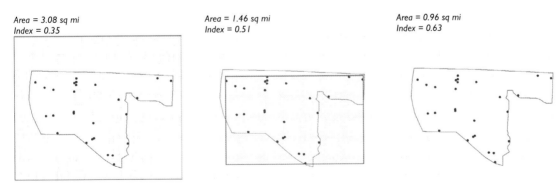

Three ways of defining the extent of the study area: the full extent of the map view on the screen (left), the envelope encompassing the point features (center), and the areal extent of a polygon (in this case, a neighborhood boundary) containing the points (right). The larger the area, the lower the nearest neighbor index.

Using the envelope encompassing the features is the least desirable alternative, since it guarantees that at least two points will be at the edge of the area, so the areal extent will likely be underestimated. If you have the option, you should specify the actual areal extent of the study area.

Where the boundaries of the study area fall can affect the results of the analysis. A feature's nearest neighbor won't be used if it's outside the boundary, so you may be missing the feature's true nearest neighbor. This often happens when the study area is a political or administrative unit.

Nearest neighbor

Emergency calls within a neighborhood (red) and calls surrounding the neighborhood (yellow). In some cases, the nearest neighbor is outside the neighborhood.

Since the nearest neighbor inside the boundary is farther away than the true nearest neighbor, the mean distance will be larger than it would otherwise be. The analysis could show a dispersed pattern that may not actually exist, since the distances used to calculate the mean distance may be larger than if the points outside the boundary were used. You may need to include surrounding areas to get a more accurate measurement of the pattern. At the very least, you'll want to map features in surrounding areas to see if there are any features near the boundary of your study area. (See "Using statistics with geographic data.")

COUNTING THE NUMBER OF FEATURES WITHIN DEFINED DISTANCES

In this method, known as the *K*-function, the GIS counts the number of neighboring features within a given distance of each feature, then sums the values. If the number of features found within the distance is greater than that for a random distribution, the distribution is clustered. The value is automatically calculated at several distances and displayed on a chart so you can see at what distance the clustering is greatest.

The *K*-function is similar to the nearest neighbor index in that it attempts to give a measure of the clustering or dispersion of the features using the distance between them. Unlike the nearest neighbor index, though, the *K*-function includes all neighbors occurring within a given distance, rather than the distance to each feature's single nearest neighbor.

The *K*-function is most often used with individual point features, such as the locations of emergency calls or bird nests.

The K-function can be used to measure patterns for point features such as the locations of emergency calls.

K refers to the distance bands that are established around each feature. Depending on the software you're using, you might specify the number of bands and the distance interval. Or, you specify the number of distance bands to use and the software calculates the distance interval. It first calculates half the maximum dimension of the study area (in either the x or y direction—whichever is greater), which becomes the maximum distance. It then divides this value by the number of distances you specified to get the interval. So, if you specify 20 distances and the study area is 10,560 feet in the east–west direction and 15,840 feet in the north–south direction, the maximum distance would be 7,920 feet (15,840/2) and the interval would be 396 feet.

Using more distances will produce a smoother curve on the chart than using fewer distances will. You'll then be able to identify more closely the distance at which the clustering is greatest. After a point, increasing the number of distances does not provide additional information.

How the *K*-function is calculated

The GIS finds the distance from each point to every other point, and then, for each point, counts up the number of surrounding points within the given distance. It's as if it had drawn a circle around a point in the set using a radius equal to the first distance, counted the number of points found within the circle, then moved to the next point, and the next, until it had summed the surrounding points for all points in the set.

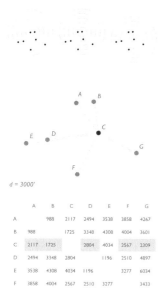

d = 3000'

	A	B	C	D	E	F	G
A		988	2117	2494	3538	3858	4267
B	988		1725	3348	4308	4004	3601
C	2117	1725		2804	4034	2567	2309
D	2494	3348	2804		1196	2510	4897
E	3538	4308	4034	1196		3277	6034
F	3858	4004	2567	2510	3277		3433
G	4267	3601	2309	4897	6034	3433	

The value of the K statistic at a given distance (d)

The distance (d) is measured between each target feature (i) and every other feature (j); each distance is multiplied by the weight for the pair (I$_{ij}$), and all these values are summed (i ≠ j means the distance between the target and itself is not included in the sum)....

$$K(d) = \frac{A}{n^2} \sum_{i \neq j} \sum I_{ij} \, d_{ij}$$

....then the result is multiplied by the area (A) divided by the number of features (n) squared

I is a weight and is either 1, if the neighboring point is within the distance of the target point, or 0 if it's not. Each neighboring point is multiplied by *I*—if it's outside the distance, the product is 0, and nothing is added to the sum. That way, only a neighboring point within the distance is counted with the target point.

The GIS calculates a *K* value at each distance, according to the number of distances you specified.

Interpreting the results of the *K*-function

To see if there's a pattern, you compare the observed *K* value at each distance to the expected *K* value for a random distribution (which is calculated by the software, given the same area and number of points). This is usually done by plotting the values on a chart, with the *K* values on the y-axis against the distances on the x-axis.

The *K* values get very large as the distance increases, so to reduce the height of the y-axis and make the chart easier to read, transformed values are plotted instead of the raw *K* values.

One method of transforming the *K* values is a variation of the *K*-function known as *L(d)*. This transformation multiplies the area by the number of point pairs for a given distance, divides the result by π times the possible number of pairs, and takes the square root.

The distance (d) is measured between each target feature (i) and every other feature (j); each distance is multiplied by the weight for the pair (I$_{ij}$), which is one if within the distance and zero if not....

The value of L at a given distance (d)

$$L\left(d\right) = \sqrt{\frac{A \sum\limits_{i \neq j}\sum I_{ij}\, d_{ij}}{\pi n \left(n - 1\right)}}$$

....the distances are summed (excluding the distance between the target and itself) and multiplied by the area (A)....

....then the result is divided by pi times the number of possible point pairs represented as the number of points (n) times the number less one; and the square root is taken.

With this method, the expected value for any distance—given a random distribution of points—is the distance *(d)* itself, producing a line at a 45° angle for the expected distribution.

At any given distance, if the line for the observed *L* values is above that for the expected values, the distribution is more clustered than expected for a random distribution. If it's below the line for the expected values, the distribution is more dispersed.

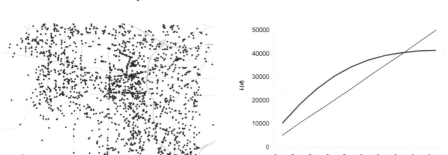

The L *values at each distance (in this case, for emergency calls) are plotted on a chart.*

Since a higher *L* value corresponds to a higher number of points found at that distance, a peak in the chart indicates clustering at a particular distance. A second peak may indicate that the clusters themselves are clustered. Conversely, a dip in the chart indicates that the distribution is less clustered at that distance.

To determine if the pattern is statistically significant, you compare the curve for the observed distribution to confidence limits for a random distribution. To create the confidence limits, the GIS randomly generates the values for the x- and y-coordinates and calculates the L values for that distribution, at each distance.

There may be a wide variation in values between any two random distributions, so the GIS generates a number of random distributions, 1,000 or more, to get a range of possible L values. The GIS selects the lowest and highest value at each distance from the simulations. It plots these to create an envelope (the lowest values at each distance as one curve and highest as another). The envelope defines the confidence limits.

An observed L value that exceeds the upper confidence limit indicates a statistically significant clustered pattern for that distance, while one that falls below the lower limit indicates a statistically significant dispersed pattern.

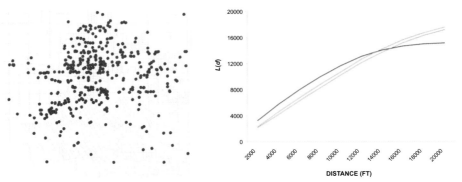

Assaults are significantly clustered up to about 2.5 miles (13,000 ft.); beyond 3 miles the pattern is significantly dispersed.

Using the *K*-function to compare the observed distribution to other distributions

You can compare distributions by calculating the *L* values for the distributions and plotting them on a chart. For example, you might compare the distributions of burglaries, assaults, and auto thefts to each other, to see if one is more clustered than the others, at any given distance.

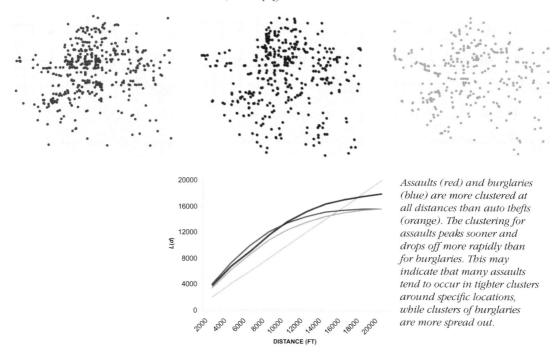

Assaults (red) and burglaries (blue) are more clustered at all distances than auto thefts (orange). The clustering for assaults peaks sooner and drops off more rapidly than for burglaries. This may indicate that many assaults tend to occur in tighter clusters around specific locations, while clusters of burglaries are more spread out.

You can also compare the observed distribution—such as the distribution of crimes—to the distribution of control data—such as the distribution of the population of a city. As CrimeStat author Ned Levine points out, this is particularly useful for social phenomena, which are often not random. You could, for example, determine whether the distribution of emergency calls in a city is more clustered than the distribution of the population as a whole by calculating the *L* values using census block centroids (the *L* values are in essence weighted by the number of people in each block—if there are 275 people in a block, the GIS calculates the *L* value as if there were 275 features at that location). This would be more useful than comparing the emergency calls to a hypothetical random distribution, because the calls might tend to cluster simply because the underlying population is similarly clustered. (See "Using statistics with geographic data.")

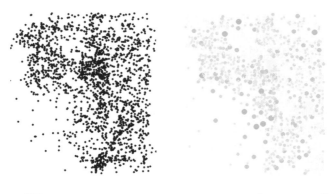

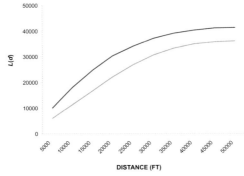

Emergency calls (red) are slightly more clustered than the population in general (represented by census block centroids, shown in orange—the larger and darker the circle, the greater the population for that block). This indicates that something other than the distribution of population is causing the pattern.

Factors influencing the *K*-function results

Points near the edge of the study area are likely to have fewer neighboring points within the distance because some neighboring points might be outside the study area. This is often the case if you're using an administrative area, such as a county, as the boundary of the study area.

At larger distances it's more likely that fewer neighboring points will be found than actually exist, since these distances are more likely to extend beyond the boundary of the study area, so it's best to be skeptical of any patterns that appear at larger distances. Depending on the software you're using, you may have the option of calculating a *K*-function statistic that corrects for this effect. (See "Using statistics with geographic data.")

COMPARING METHODS FOR MEASURING THE PATTERN OF FEATURE LOCATIONS

Overlaying areas of equal size	Quadrat analysis	Kolmogorov-Smirnov Chi-square Variance-mean ratio	Can be used when there are multiple features at a single location	Doesn't consider the distance between features; results are influenced by the size of the quadrats
Calculating the average distance between features	Nearest neighbor index	Z-score	Considers the distance between features	Results may be biased if there are many features near edge of study area
Counting the number of features within defined distances	K-function	Uses multiple simulations to create a random distribution envelope	Calculates the concentration of features at a range of scales or distances, simultaneously	Patterns are suspect at larger distances due to edge effects

In addition to measuring the pattern formed by the locations of features, you can also measure patterns of attribute values associated with features, such as the pattern formed by median house values. These methods reveal whether similar values tend to occur near each other, or whether high and low values are interspersed.

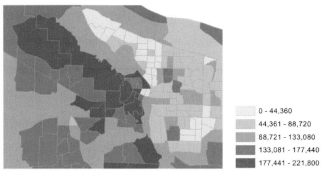

	0 - 44,360
	44,361 - 88,720
	88,721 - 133,080
	133,081 - 177,440
	177,441 - 221,800

Median house value by census tract.

The idea behind measuring patterns of feature values

Measuring the spatial pattern of feature values is based on the notion that things near each other are more alike than things far apart, an idea often attributed to geographer Waldo Tobler. The idea is consistent with our everyday experience; for example, the climate where you live is likely to be more similar to the climate in nearby areas than to the climate far to the north or south of you; the assessed value of a house tends to be similar to those of surrounding houses.

Similar values for nearby features often occur because of similar conditions. For example, high crop yields for neighboring farms in an area likely result from the combination of fertile soil, high soil moisture, and high soil nutrient content in that location. Sometimes, features having similar values simply attract each other, regardless of other conditions at the location. For example, ethnic communities within cities, often established when a few immigrants located in a particular neighborhood, attract relatives and others of the same ethnicity, eventually resulting in the area having a high proportion of people of that group.

Of course, there are exceptions to Tobler's principle. A city may have a climate more similar to another city on the same side of a mountain range—even if that city is far away—than to a closer city on the other side of the range. Similarly, neighborhoods can change abruptly if separated by a highway or river. In general, though, the principle holds true for both natural and social phenomena, especially over large areas and over longer periods of time.

Statisticians call this phenomenon spatial autocorrelation. Spatial auto-correlation indicates whether the distribution of *values* is dependent on the spatial distribution of the *features*—that is, whether particular values are likely to occur in one location, or are equally likely to occur at any location.

If nearby features are more like each other than they are like more distant features, there is said to be positive spatial autocorrelation. If neighboring features tend to be unlike each other, this is termed negative spatial auto-correlation. In a random pattern there is no spatial autocorrelation. For geographic phenomena, positive spatial autocorrelation is more common.

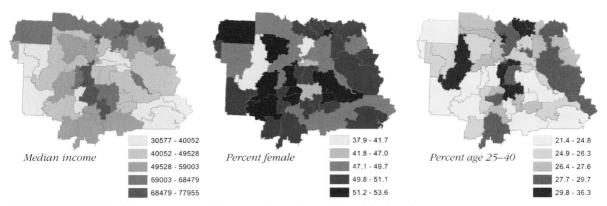

Median income	30577 - 40052
	40052 - 49528
	49528 - 59003
	59003 - 68479
	68479 - 77955

Percent female	37.9 - 41.7
	41.8 - 47.0
	47.1 - 49.7
	49.8 - 51.1
	51.2 - 53.6

Percent age 25–40	21.4 - 24.8
	24.9 - 26.3
	26.4 - 27.6
	27.7 - 29.7
	29.8 - 36.3

The dataset on the left exhibits positive spatial autocorrelation (ZIP Codes with high median income are found near each other), while the one in the middle exhibits negative spatial autocorrelation. The distribution on the right is random—there is no spatial autocorrelation.

All methods for measuring the spatial pattern of values produce a single statistic that summarizes both the difference in attribute values between features and how far apart the features are. The statistic indicates whether there is positive spatial autocorrelation, negative, or none—that is, whether the feature values are clustered, dispersed, or randomly distributed.

Finding patterns for areas with categories

This method, referred to as the join count statistic, is used with contigu-ous areas having category attributes (nominal data), such as parcels classi-fied as either developed or vacant. The GIS finds all shared borders (joins) between areas and counts the number of these for which the value on either side of the border is the same and the number for which the values are different.

The number of joins of each type is compared with the expected number based on the probability of two adjacent areas having the same value by chance. If the counted number of joins for areas having the same value is greater than the expected number, that value is clustered.

It is often used with areas having one of only two possible categories (either/or, or binary data), although it can be used with areas having any of several possible categories (multinomial data). You'd use the join count statistic, for example, when analyzing the pattern of election results by county in a two-person race. The information would be useful for candidates in future elections, to know if they will need to advertise widely or can concentrate on smaller areas. The join count statistic could be used to measure the pattern of counties that have had one or more cases of a particular disease, such as West Nile virus, and those that haven't. Knowing this will tell you if you can concentrate your resources to combat the disease, or if you will need to disperse them across the region.

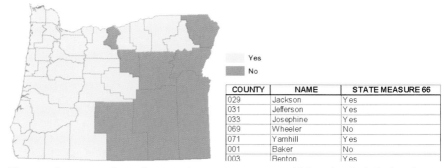

COUNTY	NAME	STATE MEASURE 66
029	Jackson	Yes
031	Jefferson	Yes
033	Josephine	Yes
069	Wheeler	No
071	Yamhill	Yes
001	Baker	No
003	Benton	Yes

The vote—by county—for and against Oregon State Measure 66 (1998), which dealt with funding for parks and open space

Finding patterns for features having continuous values

These methods are used for features having interval or ratio values, such as the percentage of the population age 65 and over in each block group. The analysis compares the attribute values between neighboring features to the distribution of values for the dataset as a whole. The methods can be used with any type of geographic features (discrete features, spatially continuous data, or contiguous areas containing summarized data).

High : 0.258

Low : 0.0424

Raster surface of maximum ozone readings (over an eight-hour period)

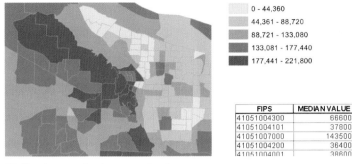

FIPS	MEDIAN VALUE
41051004300	66600
41051004101	37800
41051007000	143500
41051004200	36400
41051004001	38600

Median house value by census tract

Measuring the similarity of nearby features

Two statistics are commonly used. One uses the difference in values between any two neighboring features. It was developed by economist and statistician Robert Geary in the early 1950s and is termed Geary's contiguity ratio (or Geary's c). Geary first applied it to identify agricultural and demographic patterns in his native Ireland.

The other statistic doesn't compare the attribute values for neighboring features directly to each other, but compares the value for each feature in the pair to the mean value for the dataset. This statistic is termed Moran's Index (or Moran's I) after Australian statistician Patrick Moran, who developed it in the late 1940s.

For both methods, if the difference in values of nearby features is less than the difference in values among all features, like values are clustered. The methods only indicate that similar values occur together—you don't know from either statistic whether any clusters are composed of high values or low values.

A county public health agency deciding how to provide health services would want to know whether census tracts having similar median household income are clustered or dispersed. If there is clustering, the agency could establish clinics in low-income areas. If low-income areas are dispersed, the agency could use mobile services, such as visiting nurses, to provide health care.

Measuring the concentration of high or low values

This method is also used with continuous values (interval or ratio data). It measures within a specified distance how high or low the values are, and compares this to a measure of how high or low the values are over the entire study area. The method, developed by regional scientist Art Getis and statistician Keith Ord in the early 1990s, is known as the General G-statistic—General since it calculates a single statistic for the entire study area. (A local version that calculates a statistic for each feature in the study area is discussed in chapter 4, "Identifying clusters.")

The General G-statistic measures concentrations of high or low values over the entire study area. You might use it to compare the pattern of different types of crimes in a city to see which ones occur in hot spots and which are dispersed. Knowing this would help determine an effective approach to preventing the crimes. Unlike Geary's c and Moran's I, the General G-statistic tells you whether high or low values are concentrated.

Comparing methods for measuring the pattern of feature values

Join count	Categories	Whether values are clustered or dispersed	Straightforward way to identify patterns for areas	Only applies to categorical (nominal) data
Geary's c Moran's I	Continuous	Similarity of nearby features	Provides a single statistic summarizing the pattern	Doesn't indicate if clustering is for high values or low values
General G	Continuous	Concentration of high/low values	Indicates whether high or low values are clustered	Wor s best when either high or low values cluster (but not both)

FINDING PATTERNS FOR AREAS WITH CATEGORIES

The join count statistic is often used with either/or (binary) attribute values. For example, a planner might want to know if vacant land parcels are clustered within a county. To simplify the analysis, the planner could reclassify the typical land-use type classes of commercial, industrial, residential, vacant, and so on, as either "developed" or "vacant." He could then use the join count statistic to produce a measure of the extent to which vacant parcels cluster or the extent to which they are interspersed with developed parcels. If vacant parcels are clustered, the planner could propose large developments on adjacent parcels. If vacant parcels are dispersed, smaller projects would be called for.

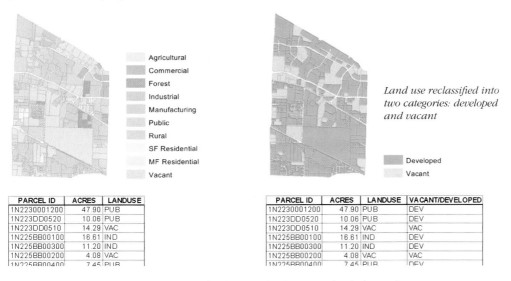

Agricultural
Commercial
Forest
Industrial
Manufacturing
Public
Rural
SF Residential
MF Residential
Vacant

Land use reclassified into two categories: developed and vacant

Developed
Vacant

PARCEL ID	ACRES	LANDUSE
1N2230001200	47.90	PUB
1N223DD0520	10.06	PUB
1N223DD0510	14.29	VAC
1N225BB00100	16.61	IND
1N225BB00300	11.20	IND
1N225BB00200	4.08	VAC
1N225BB00400	7.45	PUB

PARCEL ID	ACRES	LANDUSE	VACANT/DEVELOPED
1N2230001200	47.90	PUB	DEV
1N223DD0520	10.06	PUB	DEV
1N223DD0510	14.29	VAC	VAC
1N225BB00100	16.61	IND	DEV
1N225BB00300	11.20	IND	DEV
1N225BB00200	4.08	VAC	VAC
1N225BB00400	7.45	PUB	DEV

A conservation biologist wants to find out which watersheds have large blocks of logged or unlogged areas, as opposed to having small logged areas interspersed among unlogged areas. Areas having large unlogged blocks would provide habitat for more species and larger animals. She would first classify the subwatersheds within each major watershed as either logged or not logged, using some threshold value (for example, less than 10% logged might be classified as unlogged). She could then use the join count method to get an overall measure of whether, for each watershed, logged and unlogged areas are clustered or interspersed. Some watersheds may be either mostly logged or mostly unlogged; some may have logged areas in one part only; and some may have logged and unlogged areas interspersed. The ones with clustered unlogged areas could become the core of a conservation area.

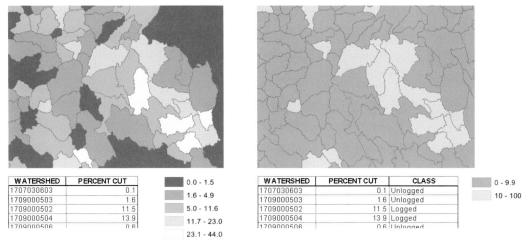

WATERSHED	PERCENT CUT
1707030603	0.1
1709000503	1.6
1709000502	11.5
1709000504	13.9
1709000506	0.6

- ■ 0.0 - 1.5
- ■ 1.6 - 4.9
- ■ 5.0 - 11.6
- □ 11.7 - 23.0
- □ 23.1 - 44.0

WATERSHED	PERCENT CUT	CLASS
1707030603	0.1	Unlogged
1709000503	1.6	Unlogged
1709000502	11.5	Logged
1709000504	13.9	Logged
1709000506	0.6	Unlogged

- ■ 0 - 9.9
- □ 10 - 100

Watersheds by percent cut (left) and reclassified into two categories: logged (more than 10% cut area) and unlogged

Join count analysis doesn't take into account the magnitude of the value in each area since it is based simply on whether the area falls into one category or the other. For example, one county may have voted 51% "Yes" for a state measure and 49% "No," while the adjacent county voted 70% "Yes" and 30% "No." Join count only considers the fact that both counties are in the "Yes" column. The other methods for measuring patterns do consider the magnitude of the values. With those methods, the clustering would be stronger if adjacent counties voted heavily "Yes."

- □ Yes
- ■ No

The vote on Oregon State Measure 66, by county. Join count does not consider the percentage of the vote, only whether the county voted for or against.

While you can reclassify continuous attribute values into two categories and perform this analysis, you may lose information in the process. Reclassifying works best with values for which there is a definite break point between the categories. For example, if you're looking for patterns in the median income of census tracts, you might compare the local median income of census tracts to the statewide median income, and assign tracts to categories for values either above or below the statewide median.

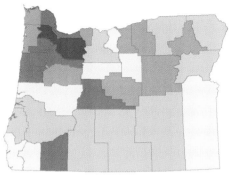

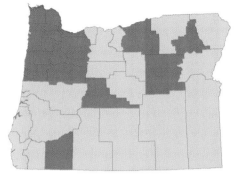

COUNTY	MEDIAN INCOME
033	31399
069	30200
071	39951
001	32786
003	41556
005	46236

☐	30,125 - 31,927
☐	31,928 - 34,063
☐	34,064 - 36,294
☐	36,295 - 41,556
■	41,557 - 51,636

COUNTY	MEDIAN INCOME	ABOVE/BELOW
033	31399	below
069	30200	below
071	39951	above
001	32786	below
003	41556	above
005	46236	above

■	Above
☐	Below

Median income by county (left), and median income reclassified to show counties above and below the median for the state as a whole

What the join count statistic measures

Since there are only two possible attribute values for the areas, you can count the number of each of three types of joins. For example, if the two possible values are "developed" and "vacant," the possible combinations are:

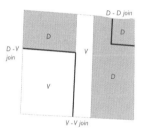

- developed-developed (adjacent areas are both developed)
- vacant-vacant (adjacent areas are both vacant)
- developed-vacant (one area is developed and the adjacent area is vacant)

The number of joins of each type, and the relative abundance of areas having either attribute value, determines whether or not there is clustering.

You can calculate the statistic using a neighborhood that includes only adjacent areas that have a shared border, or one that also includes areas that join at a corner. Defining a neighborhood is discussed in "Defining spatial neighborhoods and weights."

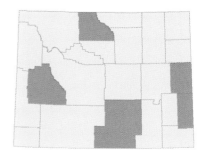

The vote by county in Oregon displays a highly clustered pattern, while the vote in Wyoming displays a dispersed pattern.

Calculating the join count statistic

If you were calculating the statistic manually, you'd make a list of all the areas and look at the adjacent areas to see how many had the same value as the first area and how many had a different value. You would ignore any areas that weren't adjacent. The GIS can't visualize like a human, however. So, for each area, it needs to look at all the other areas in the set, calculate whether each has the same or a different value, figure out which ones are adjacent, and then use a weight value of zero to exclude the non-adjacent areas from being summed for the target area. While it may use some computational shortcuts, the software essentially repeats this process for each feature.

The GIS starts by temporarily assigning 1 to one attribute value and 0 to the other, then calculates the number of 1-to-1, 0-to-0, and 1-to-0 or 0-to-1 joins.

The GIS multiplies the two attribute values and then multiplies this product by the weight assigned to that pair.

So, any 1-to-1 joins end up with a value of 1, which is then multiplied by the weight: if the area is adjacent, the weight is 1 and the resulting value will be 1. If the area is not adjacent, the weight will be 0, and the resulting value will be 0.

Any 1-to-0, 0-to-1, or 0-to-0 joins end up with a value of 0. These values are also multiplied by the weight, but since the product is 0, the result is still 0.

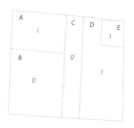

	A	B	C	D	E
A		1	1	0	0
B	1		1	0	0
C	1	1		1	0
D	0	0	1		1
E	0	0	0	1	

Pair	Values	Weight	Result
A & B	(1 * 0)	* 1	= 0
A & C	(1 * 0)	* 1	= 0
A & D	(1 * 1)	* 0	= 0
A & E	(1 * 1)	* 0	= 0

Once the values for all the pairs are calculated, they are summed. The GIS ends up counting each join twice: once with one area as the target and once with the other. So the sum of all the values is divided by two (or multiplied by ½) to get back to the correct number of joins.

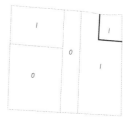

The number of 1-1 joins

The values of the target feature (x) and each other feature (x) are multiplied, and the product multiplied by the weight for that pair (w$_{ij}$)....

$$O_1 \ = \ \tfrac{1}{2} \sum_i \sum_j \left[w_{ij} \left(x_i x_j \right) \right]$$

....then the results are summed for all pairs, and divided by 2

To calculate the number of 0-to-0 joins, the software reverses the values for each pair of areas by subtracting the value of each area from 1; if the value is 0, you get 1 $(1-0 = 1)$, if the value is 1 you get 0 $(1-1 = 0)$. The rest of the calculation is the same as for calculating the number of 1-to-1 joins.

The number of 0-0 joins

The values of the target feature and each other feature are both subtracted from 1 before being multiplied

$$O_0 \ = \ \tfrac{1}{2} \sum_i \sum_j \left[w_{ij} \left(1 - x_i \right) \left(1 - x_j \right) \right]$$

To calculate the number of 1-to-0 or 0-to-1 joins, the software first subtracts one value from the other, then squares the difference (to make sure the result is positive). The product is 1 only if either of the values is 1— that is, if the values are different. If both are 1 or both are 0, the product is 0. Thus,

$1-1 = 0$, squared = 0

$0-0 = 0$, squared = 0

$1-0 = 1$, squared = 1

$0-1 = -1$, squared = 1

It then multiplies the squared value by the weight, sums these values, and multiplies by ½.

The number of 1-0 or 0-1 joins

The value of each feature is subtracted from the value of the target feature, and the difference squared before being multiplied by the weight for that pair

$$O_{01} \ = \ \tfrac{1}{2} \sum_i \sum_j \left[w_{ij} \left(x_i - x_j \right)^2 \right]$$

Testing the significance of the join count statistic results

The analysis shows you how many of each join type there are in the study area. You can test the results to see if the numbers of joins are significantly different from what you'd get in a random pattern. That will tell you if the values are clustered, and to what extent.

The test compares the observed number of joins of each type to the expected number of each type for a random distribution. Once the expected number of joins is calculated, the GIS calculates a Z-score for each type. You then need to compare the test statistics to interpret the pattern.

Calculating the expected joins for a random distribution

The expected number of joins for each type is based on the probability of an area being either one value or the other. Once the GIS knows these probabilities, it can calculate the probability of each type of join occurring, and from that the expected number of joins of each type.

You calculate the probability assuming either normalization or randomization sampling. Sampling is discussed in "Testing statistical significance."

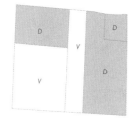

If you're using the normalization assumption, you specify the probability values you want to use. If you're using the randomization assumption, the GIS will calculate the probabilities. Assume the two possible values are developed (D) and vacant (V). The probability of a D area occurring is the number of D areas divided by the total number of areas.

The probability of a developed area occurring....

$$p_D = n_D / n$$

....equals the number of developed areas divided by the total number of areas

Once the probability of a given area being of type D is either specified or calculated, the GIS calculates the probability of a join being D-D (two adjacent D areas). This is simply the probability of one area having a value of D multiplied by the probability of an adjacent area having a value of D, which is the probability squared.

The probability of a D-D join occurring...   ....is the probability of a developed area occurring multiplied by itself....

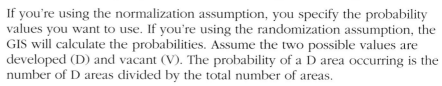

$$\text{probability of D-D join} = p_D \times p_D = p_D^2 \quad \text{....or, the probability squared}$$

Finally, the GIS multiplies the probability of a D-D join occurring by the total number of joins in the study area to get the expected number of D-D joins.

$$E_{DD} = p_D^2 L$$

The process is the same to calculate the expected number of V-V joins.

For the mixed joins (D-V or V-D), the expected number of joins is calculated using the probability of an area having a value of D and the probability of an area having a value of V (p_D and p_V). The probability of a join being mixed is the type D probability times the type V probability, doubled (since it could be either D-V or V-D).

And the expected number of D-V joins is the probability of a D-V join occurring multiplied by the total number of joins.

$$E_{DV} = 2(p_D \times p_V)L$$

Comparing the observed and expected numbers of joins

To see if the pattern is statistically significant, the software calculates a Z-score for each join type. It first calculates the standard deviation for each type. The software then calculates the Z-score by subtracting the expected number of joins from the observed number of joins, and dividing by the standard deviation. For example, for D-D joins:

$$Z_{DD} = \frac{O_{DD} - E_{DD}}{\sigma_{DD}}$$

At a confidence level of 0.05 (95%), a Z-score greater than 1.96 would indicate that areas having a value of D occur next to each other (are clustered) more than you would expect by chance. See "Testing statistical significance" for more about the Z-score.

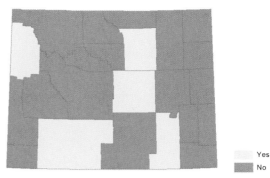

For the map on the left, the number of "No" joins is 8, the expected number is 4, and the Z-score for "No" joins is 1.29, indicating the clustering is significant at a confidence level of 0.20—you can be 80% sure the clustering isn't due to chance. In the map on the right, the number of "Yes" joins is 28, the expected number is 31, and the Z-score is −0.44, indicating a random distribution.

Factors influencing the join count statistic results

In their book, *Elementary Statistics for Geographers,* James Burt and Gerald Barber point out several situations that can make the results of the analysis suspect.

- There are fewer than 30 features in the study area. Fewer than 30 features is considered a small sample and, in general, any statistical analysis that uses a small sample is somewhat suspect simply because one extreme value can bias the results.

- One of the category values occurs in less than 20% of the areas.

- The region is elongated, so most areas have few joins.

- There are a couple of features with many joins while all other features have only one or two joins.

In these cases, you should use additional methods for analyzing the patterns. If you're interested in one of the values in particular—such as whether or not vacant parcels are clustered—you could select these features, copy them into a separate layer, and analyze them using one of the methods discussed earlier in the chapter, such as quadrat analysis or the nearest neighbor index.

Land use (left map) reclassified into "developed" and "vacant" (middle). The map on the right shows the vacant parcels selected so the pattern of parcel locations can be tested using quadrat analysis or the nearest neighbor index.

On the other hand, if the values are magnitudes (such as the percentage by county of the vote for and against a ballot measure), you could analyze the patterns using one of the methods discussed in the next sections.

Yes
No

56.5 - 59.2
53.6 - 56.4
51.4 - 53.5
47.4 - 51.3
45.6 - 47.3

Vote classified into two categories (left) and the "Yes" vote percentage in each county (right)

MEASURING THE SIMILARITY OF NEARBY FEATURES

Geary's *c* and Moran's *I* use the magnitude of feature values to identify and measure the strength of spatial patterns. A social scientist studying the integration of different ethnic groups in a region could measure the extent to which the groups are clustered or dispersed, using the percentage of the population of each group in each census tract or block group.

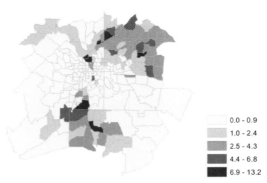

	0.0 - 0.9
	1.0 - 2.4
	2.5 - 4.3
	4.4 - 6.8
	6.9 - 13.2

Census tracts color coded by percent Asian population.

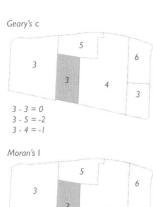

Geary's c

3 - 3 = 0
3 - 5 = -2
3 - 4 = -1

Moran's I

Mean = 24 / 6 = 4

Target   Neighbors
3 - 4 = -1   3 - 4 = -1
                5 - 4 = 1
                4 - 4 = 0

What Geary's c and Moran's I measure

Both methods compare values for neighboring features. They then compare the difference in values between each pair of neighbors to the difference in values between all features in the study area. If the average difference between neighboring features is less than between all features, the values are clustered.

Geary's *c* calculates the difference in values between the target features and each of its neighbors. Moran's *I* calculates the difference between the target feature and the mean for all features, and the difference between each neighbor and the mean. It then compares these differences for the target feature and each neighbor in turn.

For both Geary's *c* and Moran's *I*, you can specify a neighborhood based on adjacency, a set distance, or the distance to all features in the dataset. Choose according to the nature of the spatial relationship between the features you're analyzing. Defining a neighborhood is discussed in "Defining spatial neighborhoods and weights."

Calculating Geary's c

The GIS starts with one feature and subtracts the value of a neighboring feature from the value of the original, or target, feature.

Since the GIS only needs to find out how large the difference is between neighboring features, and not whether the difference is positive or negative, it squares the difference to make sure it's positive.

The GIS then multiplies the difference by a weight value (based on the neighborhood you defined). It repeats the process for the target feature with all other features in the study area. It then moves to the next feature and does the same thing, summing the results as it goes.

For each target (i) and neighbor (j) pair, the value of the neighbor (x)
is subtracted from the value of the target (x), the difference is squared,
and the result multiplied by the weight (w) for that pair....

$$\sum_i \sum_j w_{ij} \ (x_i - x_j)^2$$

....then the results are added to the sum for all pairs

Next the GIS calculates the variance—how far values are from the mean. (See "Understanding data distributions.")

Now the GIS sums the weights for each pair of features, multiplies this by two to account for each pair-wise combination (to–from and from–to), and then multiplies the variance by this result. It does this to account for the weights that were summed when calculating the difference in feature values.

The weights for all pairs are
summed, and the result doubled....

$$2 \sum_i \sum_j w_{ij} \ \frac{\sum_i (x_i - \overline{x})^2}{n}$$

....then multiplied by the variance (the difference between
each target value and the mean value squared, the result
summed for all features, and divided by the number of features)

Finally, it divides this value into the initial value it calculated (the sum of the weighted differences in values) to get the *c* ratio. The number of features *(n)* is moved to the numerator to simplify the equation.

Geary's c ratio

$$c = \frac{n \ \sum_i \sum_j w_{ij} \ (x_i - x_j)^2}{2 \ \sum_i \sum_j w_{ij} \ \sum_i (x_i - \overline{x})^2}$$

Since the weight value is used in both the numerator and denominator, the ratio is essentially the sum of the differences in attribute values between any two features within the neighborhood (as defined by the weights) compared to the sum of the differences for the dataset as a whole (as represented by the variance). Multiplying the differences by the weights incorporates the proximity of the features.

Interpreting Geary's c

A small difference in values between nearby features will result in a small numerator, indicating that similar values are more likely to be found near each other. Conversely, a large difference in values indicates that similar values are less likely to be found near each other.

While that gives you a general sense of the pattern, you can also figure out how much the value of c differs from the value you would expect for a random distribution. A value of c equal to 1 means the distribution is random (the numerator and denominator in the equation are equal).

A value of c close to 0 means the distribution of values is clustered. If all the features in the study area had close to the same value, the sum of all the pairs would be near 0, the numerator would be a very small number, and the final value for c would also be close to 0.

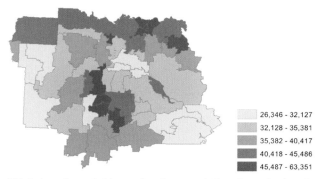

	26,346 - 32,127
	32,128 - 35,381
	35,382 - 40,417
	40,418 - 45,486
	45,487 - 63,351

ZIP Codes color coded by median income. A Geary's c value less than 1 (0.46 in this case) indicates clustering.

Conversely, a value of c greater than 1 indicates the values are dispersed. The numerator is larger than the denominator—the differences in values for neighboring features are greater than the difference between values for the study area as a whole.

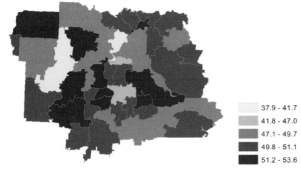

	37.9 - 41.7
	41.8 - 47.0
	47.1 - 49.7
	49.8 - 51.1
	51.2 - 53.6

ZIP Codes color coded by percent female population. A c value greater than 1 (1.26) indicates a dispersed pattern.

Calculating Moran's *I*

For each pair of features, the GIS subtracts the value of each feature from the mean value for all the features in the study area. It then multiplies these values to get what's known as the cross-product. The GIS repeats the process for all pairs of features in the study area and sums the results.

The mean is subtracted from the value of the target feature and the value of the neighbor, and the differences are multiplied....

$$\sum_i \sum_j w_{ij} \; (x_i - \overline{x})(x_j - \overline{x})$$

....then the result is multiplied by the weight for that pair, and added to the sum for all features

The software then calculates the variance from the mean value for all the features in the study area, sums the weights for each pair of features, and multiplies the variance by this sum, just as with Geary's *c*.

The variance is multiplied by the sum of the weights

$$\sum_i \sum_j w_{ij} \; \frac{\sum_i (x_i - \overline{x})^2}{n}$$

Finally, it divides this value into the initial value it calculated (the sum of the weighted cross-products) to get the ratio. Again, like Geary's *c*, the number of features variable (*n*) is moved to the numerator.

Moran's Index

$$I = \frac{n \; \sum_i \sum_j w_{ij} \; (x_i - \overline{x})(x_j - \overline{x})}{\sum_i \sum_j w_{ij} \; \sum_i (x_i - \overline{x})^2}$$

Interpreting Moran's *I*

In the equation for Moran's *I,* the GIS multiplies the difference from the mean for each pair of features—a high cross-product indicates nearby features have similar values; a low cross-product indicates nearby features have dissimilar values.

For example, if the mean is 6 and the target feature has a value of 12, the cross-product for a neighboring feature with a value of 10 is 24:

$$(12-6) \times (10-6) = 6 \times 4 = 24$$

If both neighboring values are less than the mean—say 3 and 2—the cross-product is still positive, but is less than the cross-product for the higher values:

$$(3-6) \times (2-6) = -3 \times -4 = 12$$

If one value of the pair is greater than the mean and one is less than the mean, then the product is negative. For example, if the target feature has a value of 12 but the neighboring feature has a value of 3, the cross-product is negative, in this case, −18:

$$(12-6) \times (3-6) = 6 \times -3 = -18$$

If there are roughly as many pairs with positive cross-products as there are with negative cross-products, the result of summing the cross-products will be close to 0. That indicates a random distribution—some neighbors have similar values and some don't. So if you get a value for *I* that's close to 0, the distribution of feature values resembles a random distribution.

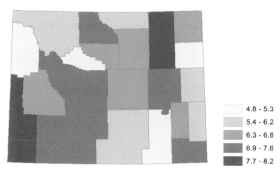

	4.8 - 5.3
	5.4 - 6.2
	6.3 - 6.8
	6.9 - 7.6
	7.7 - 8.2

Percent of population 5 to 10 years of age, by county. An I of −0.06 (close to 0) indicates the distribution is similar to a random distribution.

You can calculate the expected value of *I* if the values are distributed randomly, using the number of features in the study area.

The expected Moran's I for a random distribution

$$I_E = \frac{-1}{n-1}$$

-I is divided by the number of features (n), minus I

This turns out to be a small negative number, close to 0. So, for all practical purposes, a value of 0 for *I* is used to indicate a random distribution.

If more pairs have similar values than not, the sum of the cross-products will be positive, and *I* will be greater than 0. That means similar values are clustered.

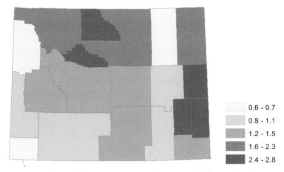

0.6 - 0.7
0.8 - 1.1
1.2 - 1.5
1.6 - 2.3
2.4 - 2.8

Percent of population 85 years of age and above. An I *greater than 0 (0.26, in this case) indicates a clustered pattern.*

If there are more pairs with dissimilar values, the sum of the cross-products will be negative, and *I* will be less than 0. That means the values are dispersed.

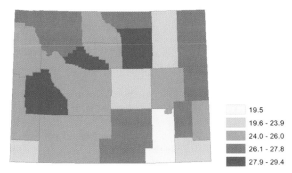

19.5
19.6 - 23.9
24.0 - 26.0
26.1 - 27.8
27.9 - 29.4

Percent of population 45 to 65 years of age. An I *of −0.12 indicates a slightly dispersed pattern since it's a little less than 0.*

The range of possible values of *I* is −1 to 1. If all neighboring features had close to the same value, the equation would calculate to near 1, indicating complete clustering of values. Conversely, if the values are completely dispersed, the value of *I* calculates out to a value near −1.

This table, based on one by Michael Goodchild, shows the possible values of Geary's c and Moran's I, and the pattern they indicate.

$c < 1$	$I > 0$	Clustered (similar values are found together)
$c = 1$	$I = 0$	Random (no apparent pattern)
$c > 1$	$I < 0$	Dispersed (high and low values are interspersed)

Testing the significance of the results for Geary's c and Moran's I
For both Geary's c and Moran's I, the GIS calculates a Z-score to indicate how confident you can be that any pattern is not simply due to chance.

The GIS calculates the expected statistic for a random distribution, calculates the variance for that statistic, subtracts the expected statistic from the observed statistic, and divides by the square root of the variance (the standard deviation).

The Z-score for Geary's c

The expected c is subtracted from the observed c....

$$Z_c = \frac{c_O - c_E}{SD_{c_E}}$$

....and the difference divided by the standard deviation of c for the expected distribution

$$Z_I = \frac{I_O - I_E}{SD_{I_E}}$$

The Z-score for Moran's I is calculated the same way

While the expected value of Geary's c for a random distribution is always 1, the expected value of Moran's I for a random distribution depends on the number of features in the study area, calculated as $-1/(n-1)$, where n is the number of features. This is a very small number, close to 0.

The calculation of the variance is based on the number of features in the study area, the number of neighboring features, and the sum of the weight values. The equation for the variance (and thus the Z-score value) depends on the sampling assumption you use—either randomization or normalization.

Some software asks you to specify the sampling assumption, while other software calculates the Z-score using both assumptions, or simply chooses one (in the latter case you should check the software documentation to see which assumption is used). Sampling is discussed in the section "Testing statistical significance."

For Geary's *c,* a positive Z-score indicates a dispersed pattern, while a negative Z-score indicates clustering. Since 1 (the expected value for a random distribution) is subtracted from the observed *c* to calculate the numerator of the equation, a *c* value less than 1 (indicating clustering) will result in a negative number for the numerator, and thus a negative value for Z. Conversely, a positive *c* value (indicating a dispersed pattern) will result in a positive number for the numerator, and a positive value for Z.

For Moran's *I,* the opposite is true. Since the expected *I* is close to 0, when it is subtracted from the observed *I,* an observed *I* greater than 0 (indicating clustering) will result in a positive numerator, and a positive Z-score. An observed *I* less than 0 (indicating a dispersed distribution) will result in a negative numerator and a negative Z-score. So, a positive Z-score indicates clustering, while a negative Z-score indicates a dispersed pattern. See "Testing statistical significance" for more about the Z-score.

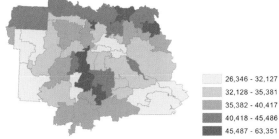

	26,346 - 32,127
	32,128 - 35,381
	35,382 - 40,417
	40,418 - 45,486
	45,487 - 63,351

Median income by ZIP Code. Both Moran's I and Geary's c indicate clustering. The Z-score for Moran's I is 4.34, and for Geary's c is −4.95. Those values exceed the critical value of 2.58 at a confidence level of 0.01 (−2.58 for Geary's c), so using either measure you can be 99% sure the clustering didn't occur by chance.

Using Geary's *c* or Moran's *I* to compare distributions

If you find there is a statistically significant pattern, you may want to see if there is an obvious reason for the pattern. For example, if you're analyzing the distribution of senior citizens by ZIP Code, and find that they tend to live in clusters, you may also want to analyze the distribution of the population as a whole. It may be that the senior population is clustered simply because the population in general occurs in clusters across the region. You'd do this by analyzing the population by ZIP Code, using the same statistic, and calculating the Z-score. If the Z-score for the senior population is higher (using Moran's *I*) than that for the population as a whole, you can conclude the senior population is more clustered than you would expect based on the distribution of the general population. (See "Using statistics with geographic data.")

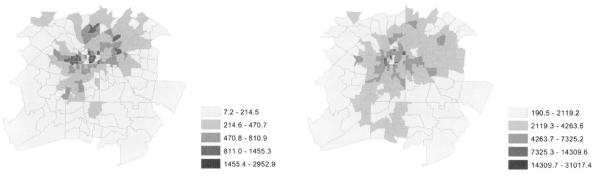

7.2 - 214.5	190.5 - 2119.2
214.6 - 470.7	2119.3 - 4263.6
470.8 - 810.9	4263.7 - 7325.2
811.0 - 1455.3	7325.3 - 14309.6
1455.4 - 2952.9	14309.7 - 31017.4

These density maps (number of people per square mile) show that with an I of 0.3 and a Z-score of 8.14, the population 65 years of age and above (left) is slightly more clustered than the population in general (right), with an I of 0.26 and a Z-score of 4.39.

Factors influencing the results of Geary's c and Moran's I

The results of Geary's c and Moran's I are affected by geographic scale and the extent of the study area. A study area may have few features, the ones with similar values occurring near each other. However, since there are few features, areas with dissimilar values could also be within the neighborhood, or even adjacent, so the statistic will not show a clustered pattern. If you expanded the study area to include more features, the statistic would show the clustering of values.

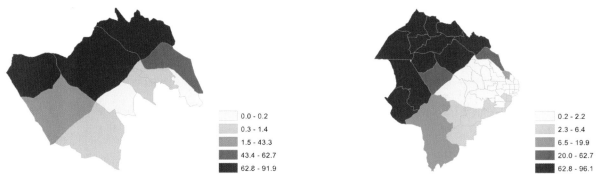

0.0 - 0.2	0.2 - 2.2
0.3 - 1.4	2.3 - 6.4
1.5 - 43.3	6.5 - 19.9
43.4 - 62.7	20.0 - 62.7
62.8 - 91.9	62.8 - 96.1

Two study areas showing percent white population by census tract. Even though the values appear to be clustered in the map on the left, Moran's I indicates no pattern—because there are so few features, the features with high values are adjacent to features with low values. In the area on the right, Moran's I indicates very strong clustering, with an I of 0.76 and a Z-score of 9.5.

MEASURING THE CONCENTRATION OF HIGH OR LOW VALUES

Geary's c and Moran's I measure whether similar values are clustered or dispersed. If you want to know whether high or low values cluster, use the General G-statistic, which measures the concentration of values over an area.

What the G-statistic measures

The G-statistic tells you whether either hot spots (clusters of high values) or cold spots (clusters of low values) exist in a study area. However, you don't know where the concentration of values is—only that across the study area, the high (or low) values tend to occur near each other. An entrepreneur opening a chain of athletic clothing stores in a region would use the G-statistic to see if there are concentrations of people likely to spend money on sporting goods, then start to look at specific locations for the stores.

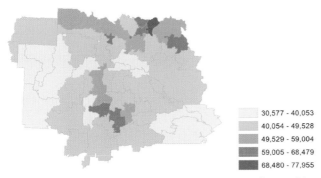

	30,577 - 40,053
	40,054 - 49,528
	49,529 - 59,004
	59,005 - 68,479
	68,480 - 77,955

Median income by ZIP Code. The General G-statistic tells you if there are clusters of ZIP Codes having high values for this attribute.

The G-statistic uses a neighborhood based on a distance you specify. Feature pairs for which the neighboring feature is within the distance of the target feature are assigned a weight of 1; all other pairs are assigned a weight of 0 (see "Defining spatial neighborhoods and weights").

Calculating the G-statistic

The GIS multiplies the attribute values for the first feature pair, then multiplies this product by the weight—1 if the features are within the distance you specified, or 0 if they aren't. It then does the same for all other pairs of features in the dataset and sums the results. For pairs where the distance is greater than the specified distance, the value ends up being 0 (since the weight is 0). Those pairs count for nothing. Finally, the sum is divided by the sum of the products of all feature pairs in the dataset.

The value of the target feature (x) is multiplied by the value of each neighbor (x); the product is multiplied by the weight for that pair (w), and added to the sum....

The General G-statistic, for a distance (d)

$$G(d) = \frac{\sum_i \sum_j w_{ij}\,(x_i \cdot x_j)}{\sum_i \sum_j (x_i \cdot x_j)}$$

....then the weighted sum is divided by the unweighted sum of the products of all features

Interpreting the results of the G-statistic

A large value for G indicates that high attribute values are found together, while a small value for G indicates that low values are found together. If you used a distance that included all the features in the dataset, the numerator and denominator would be equal, and G would be 1. In most cases, though, the distance you use will be substantially smaller. Fewer pairs will be included in the numerator, and G will be less than 1. If the pairs within the distance have relatively high values, the numerator, and hence the final value of G, will be larger.

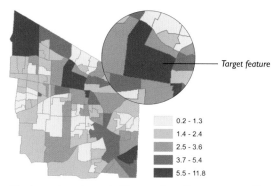

— Target feature

	0.2 - 1.3
	1.4 - 2.4
	2.5 - 3.6
	3.7 - 5.4
	5.5 - 11.8

Block groups color coded by number of emergency calls per 100 people. In the zoomed-in area, values surrounding the target feature are relatively high, adding a large number to the sum of values.

If the pairs within the distance have relatively low values, the numerator—and the value of G—will be smaller.

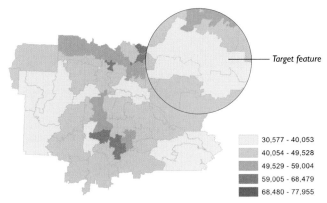

Target feature

	30,577 - 40,053
	40,054 - 49,528
	49,529 - 59,004
	59,005 - 68,479
	68,480 - 77,955

Median income by ZIP Code. The values surrounding the target feature are low, resulting in a small number being added to the sum.

Since the G-statistic is a relative measure, you don't really know what a large or small value means unless you compare it to the expected G-statistic for a random distribution at the distance you specified. You can then tell if the distribution of values is significantly different than a random distribution.

The expected G-statistic at the given distance is what the value of G would be were there no particular concentration of high or low values. It's the ratio of pairs within the distance to the total number of pairs, regardless of value. The GIS calculates this by summing the weights for the pairs and dividing by $n(n-1)$ where n is the number of features in the study area.

The expected G-statistic at a given distance, for a random distribution

$$G_E(d) = \frac{W}{n(n-1)}$$

The sum of the weights (W)....

....is divided by the number of features (n) times the number of features minus one

Since the weight is 1 for any pair within the distance, W is equal to the number of pairs within the distance. And the denominator is the total number of pairs in the study area.

If the observed G-statistic is larger than the expected G-statistic, then there is a concentration of high values.

This dataset of emergency calls per 100 people has an observed G of 0.07, higher than the expected G of 0.05, indicating that high values are clustered.

	0.2 - 1.3
	1.4 - 2.4
	2.5 - 3.6
	3.7 - 5.4
	5.5 - 11.8

If the observed G-statistic is smaller than the expected G-statistic, there is a concentration of low values.

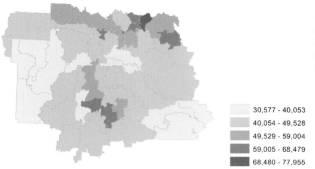

This dataset of median income has an observed G of 0.09, slightly lower than the expected G of 0.10, indicating low values are clustered.

	30,577 - 40,053
	40,054 - 49,528
	49,529 - 59,004
	59,005 - 68,479
	68,480 - 77,955

Testing the significance of the G-statistic

Once the GIS has calculated the observed and expected values of G, you can test whether the observed G is significantly different than the expected G (that is, significantly different than a random distribution) at a given confidence level.

The test involves calculating a Z-score. First the GIS calculates the variance for the expected G. The variance represents the average amount that values differ from the mean value for all the features in the study area. Then the expected G is subtracted from the observed G and divided by the square root of the variance (that is, the standard deviation).

The Z-score for the G-statistic at distance (d)

The expected G value at the distance is subtracted from the observed G....

$$Z_{G(d)} = \frac{G(d)_O - G(d)_E}{SD_{G(d)}}$$

....and the difference divided by the standard deviation for the expected G for that distance

If the observed G is larger than the expected G (that is, high values are clustered), the numerator is positive, and the Z-score is positive. If the observed G is less than the expected G (low values are clustered), the numerator is negative, and the Z-score is negative.

For example, a negative Z-score below the significant value (−1.96 at a confidence level of 95%) indicates that low values tend to be found together and you can be 95% sure the pattern is not due to chance. See the section "Testing statistical significance" for more on the Z-score.

	0.2 - 1.3
	1.4 - 2.4
	2.5 - 3.6
	3.7 - 5.4
	5.5 - 11.8

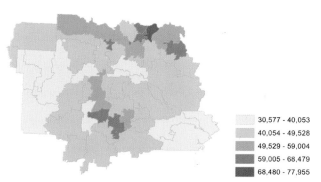

	30,577 - 40,053
	40,054 - 49,528
	49,529 - 59,004
	59,005 - 68,479
	68,480 - 77,955

With a G-statistic value higher than expected for a random distribution, and a Z-score of 4.21, the clustering of high values is significant at a confidence level of 0.01 (99%).

With a G-statistic value lower than the expected G for a random distribution (indicating clustering of low values), but a Z-score of −1.26, the clustering is not statistically significant.

Factors influencing the G-statistic results
The distance you specify when defining the neighborhood will affect the results of the analysis. If, for example, there happen to be more features with high values found within the distance, the statistic will indicate clustering of high values, even though low values may also be clustered.

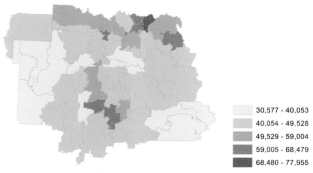

	30,577 - 40,053
	40,054 - 49,528
	49,529 - 59,004
	59,005 - 68,479
	68,480 - 77,955

At a distance of a half-mile, the G-statistic indicates clustering of low median income values, with a Z-score of −1.39, significant at a confidence level of 0.2 or 80%. At a distance of 3 miles, the Z-score is −0.43, indicating no clustering.

The size of the features you're analyzing can also affect the results. For example, if large areas tend to have low values and smaller areas tend to have high values—and both seem to be equally concentrated—the G-statistic will show that high values are concentrated, since for the small areas there are more pairs within the specified distance.

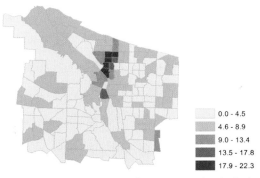

Census tracts by percentage of vacant parcels. The highest values are in the smaller census tracts, and the G-statistic shows significant clustering of high values.

The results also depend on the range of values for the features. If you have one or a few very high values—relative to the mean value for the dataset—the G-statistic may show that high values are concentrated, even if there are more features with low values near each other. This is because the equation multiplies the feature values. If there are very high values, the numerator will end up being very large.

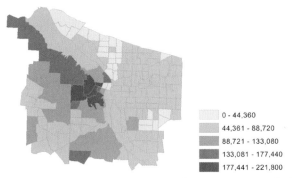

Median home value by census tract. A few tracts have very high values, causing the G-statistic to indicate significant clustering of high values.

Bailey, Trevor C., and Anthony C. Gatrell. *Interactive Spatial Data Analysis.* Longman, 1995. Includes an extensive discussion of several point pattern analysis methods, including quadrat analysis, the nearest neighbor index, and the K-function, with an emphasis on the theory and the math behind the statistics. Also includes discussions of Geary's *c* ratio and Moran's *I*.

Burt, James E., and Gerald M. Barber. *Elementary Statistics for Geographers.* Guilford Press, 1996. The chapter on correlation analysis includes a discussion of spatial autocorrelation and concise descriptions of methods to test for this, including quadrat analysis and Geary's *c* ratio, with an emphasis on the math behind the statistics.

Clark, Philip J., and Francis C. Evans. "Distance to Nearest Neighbor as a Measure of Spatial Relationships in Populations." *Ecology* 35, no. 4 (1954): 445–53. Clark and Evans describe the derivation of the nearest neighbor index.

Earickson, Robert J., and John M. Harlin. *Geographic Measurement and Quantitative Analysis.* Macmillan, 1994. The section on point pattern analysis includes brief, accessible discussions of quadrat analysis and the nearest neighbor index.

Ebdon, David. *Statistics in Geography.* Blackwell, 1985. The "Spatial Statistics" chapter includes a discussion of pattern analysis, including nearest neighbor, quadrat analysis, and Moran's *I*. The well-illustrated text also includes discussions of sampling and significance tests.

Geary, R. C. "The Contiguity Ratio and Statistical Mapping." *The Incorporated Statistician* 5 (1954): 115–45. A technical discussion of Geary's *c* ratio, with examples of its application to demographic data.

Getis, A. "Spatial Interaction and Spatial Autocorrelation: A Cross-Product Approach." *Environment and Planning* 23 (1991): 1269–77. Getis compares several methods of measuring spatial autocorrelation and describes the General G-statistic.

Getis, Arthur, and Janet Franklin. "Second Order Neighborhood Analysis of Mapped Point Patterns." *Ecology* 68, no. 3 (1987): 473–77. Getis and Franklin describe one transformation of the *K*-function, which includes edge correction, and present a forestry application.

Getis, Arthur, and J. K. Ord. "The Analysis of Spatial Association by Use of Distance Statistics." *Geographical Analysis* 24, no. 3 (1992): 189–206. Getis and Ord describe the theory behind the G-statistic and compare it to Moran's *I*. They also present two applications of the statistic.

Goodchild, Michael F. *Spatial Autocorrelation.* Catmog 47, Geo Books, 1986. A classic work on spatial autocorrelation.

Griffith, Daniel A. *Spatial Autocorrelation: A Primer.* Resource Publications in Geography, Association of American Geographers, 1987. Another classic work on spatial autocorrelation.

Lee, Jay, and David W. S. Wong. *Statistical Analysis with ArcView GIS.* Wiley, 2001. Lee and Wong's concise text covers a range of methods for analyzing patterns. The calculations and concepts are illustrated with examples.

Lembo, Arthur J. Jr. Lecture slides for Crop and Soil Science Course 622, Spatial Modeling and Analysis (unpublished), Cornell University, 2003.

Continued

References and further reading (continued)

Levine, Ned. *CrimeStat: A Spatial Statistics Program for the Analysis of Crime Incident Locations (v 2.0).* Ned Levine & Associates and the National Institute of Justice, 2002. The CrimeStat documentation covers several methods of pattern analysis, including Geary's c ratio, Moran's I, nearest neighbor index, and the K-function. Includes examples of the application of the statistics to crime analysis.

Defining spatial neighborhoods and weights

Several of the methods discussed in chapters 3 and 4 show you how to analyze patterns and clusters of feature values. These methods look at both the difference between the values of features and the spatial relationship between the features (distance or other measure).

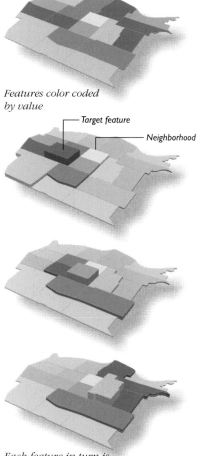

Features color coded by value

Target feature

Neighborhood

Specifically, the GIS compares the value of a feature (the "target") to the values of neighboring features. It then moves to the next feature and does the same thing, and so on, for all the features in the study area. In order to do this, the GIS requires that you define the area surrounding each target feature within which feature values are compared—termed the "neighborhood"—and the nature of the spatial relationship between features. The GIS then assigns weights to each feature pair to specify whether the two features are in each other's neighborhoods, and to represent the spatial relationship between the features.

You define the neighborhood based on the interaction between features. Features might influence each other—for example, the value of a property is often strongly influenced by the values of neighbor-

Each feature in turn is compared to its neighbors.

ing properties. Or, there may be movement between features that causes the values of nearby features to be similar. For example, where there is an outbreak of a contagious disease in a particular census tract, nearby census tracts will also have high incidence of the disease as people move to and from the surrounding areas, infecting others. Or it might be that there are underlying factors resulting in nearby features having similar values. For example, two adjacent counties may have voted heavily for the same candidate because they are both in a farming area and the people living there share the same opinions as the candidate on issues related to agriculture.

In most cases, these spatial relationships are represented using adjacency (contiguous features that share a border) or Euclidean distance. Adjacency-based neighborhoods are often used for contiguous areas or rasters, while distance-based neighborhoods can be used for discrete or contiguous features.

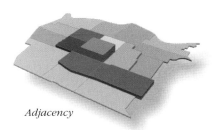

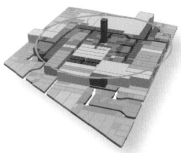

Adjacency

Distance: discrete features   *Distance: contiguous features*

You could define a neighborhood using other measures, such as travel time, travel cost, measures of attraction between locations (such as the attraction of various resorts to cities), or any other measure of spatial relationships. However, these measures can be hard to obtain, or even to quantify.

The GIS implements a neighborhood using weight values. The GIS first determines the relationship between each pair of features in the study area—using the type of neighborhood you define—and assigns a weight value to that pair. The value is stored in a table (known as a spatial weights matrix) with a row and a column for each feature, populated by the weight for each feature pair. When the GIS calculates a particular statistic, the GIS looks up the weight in the matrix.

Some GIS software creates the matrix on the fly when you perform an analysis. It either uses a default neighborhood that is appropriate for the particular analysis method, or it prompts you for the type of neighborhood if there are several different neighborhood types that would be appropriate. Other GIS software allows you to create a weights matrix as a separate step before using a particular statistical tool.

For adjacency-based neighborhoods, a weight of 1 means the nearby feature is in the neighborhood. A weight of 0 indicates the feature is not adjacent and therefore not in the neighborhood.

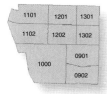

Weights matrix for an adjacency-based neighborhood

	1101	1201	1301	1102	1202	1302	1000	0901	0902
1101	0	1	0	1	1	0	0	0	0
1201	1	0	1	1	1	1	0	0	0
1301	0	1	0	0	1	1	0	0	0
1102	1	1	0	0	1	0	1	0	0
1202	1	1	1	1	0	1	1	1	0
1302	0	1	1	0	1	0	0	1	0
1000	0	0	0	1	1	0	0	1	1
0901	0	0	0	0	1	1	1	0	1
0902	0	0	0	0	0	0	1	1	0

For distance-based neighborhoods, the weight value is calculated using the distance between the features, so each pair potentially has a unique weight value.

Weights matrix for a distance-based neighborhood

	A	B	C	D	E
A	0	353	516	641	757
B	353	0	357	837	1025
C	516	357	0	659	901
D	641	837	659	0	263
E	757	1025	901	263	0

You can also create a neighborhood using a threshold or cutoff distance. Within this distance neighbors can have a weight of 1, or the weight can be calculated using the distance value.

Weights matrices using a threshold distance

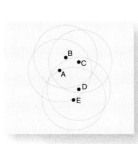

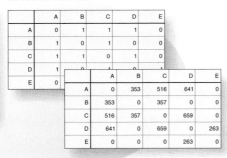

	A	B	C	D	E
A	0	1	1	1	0
B	1	0	1	0	0
C	1	1	0	1	0
D	1	0	1	0	1
E	0	0	0	1	0

	A	B	C	D	E
A	0	353	516	641	0
B	353	0	357	0	0
C	516	357	0	659	0
D	641	0	659	0	263
E	0	0	0	263	0

If the influence of feature values on surrounding features is minimal or is limited to those close by, or the phenomenon is very localized, use a neighborhood based on adjacency. For example, if you're analyzing the vote for a particular candidate in an election by county (using that candidate's percentage of the vote as the attribute value), you'd use a neighborhood based on adjacent counties. In most places, the vote is localized, with urban counties voting differently than rural counties.

If distant features exert some influence on each other or have some interaction, even if minimal, you would use a neighborhood that includes all features in the study area, with decreasing influence as the distance increases. You'd use this type of neighborhood, for example, when analyzing the pattern of occurrence of a contagious disease. You can assume that people may travel (and infect others) throughout the study area, but that they are more likely to come into contact with those in nearby areas.

When you're using a distance-based neighborhood with point features, the GIS measures the point-to-point distance. For lines and areas you can measure the distance between feature centroids or between the nearest locations on the features.

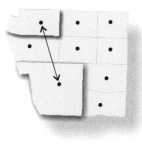

- Distance between centroids. Use this distance if the features are roughly the same size and shape.

- Distance between nearest locations on line or boundary. Use this distance if the size or shape of the features varies a great deal—otherwise, the weight may be underestimated for particularly large or convoluted features since the centroid could be far from one end of the line or the border of the area.

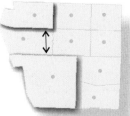

The distance can be very different depending on whether you measure from centroid-to-centroid (top) or border-to-border (bottom).

If you know that the influence of features on each other drops off completely after a certain distance, you can define the neighborhood to reflect this. This is the case, for example, with property values, where there is an accepted rule of thumb that the influence of a property on others extends to about a quarter-mile—past that distance there is no longer much influence.

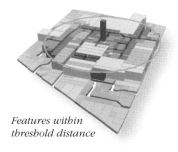

Features within threshold distance

When using a cutoff distance with point features, a single point is either within the distance or not. For features other than points, however, including lines, areas, and rasters, the feature could fall partially outside the distance. You need to decide if a feature is included if it doesn't fall completely within the distance. The options are:

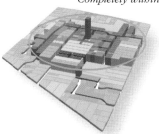

Completely within

- only features that fall completely within the distance.

- features that fall mostly within the distance. This is usually defined as the centroid of the feature falling within the distance. In some cases features are included if half the length (for line features) or more than half the area (for area features) falls within the distance.

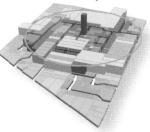

Mostly within

- features that fall partially within the distance—any part of the feature is within the distance.

If there is interaction between the features, you might want to include features that fall partially or mostly within the distance, to account for that interaction. For example, if you're analyzing earthquake faults by mean magnitude to identify clusters of faults that have had earthquakes of similar strength, you'd include faults that fall mostly within the cutoff distance, since earthquakes on one fault can trigger quakes on nearby ones. You'd want to exclude faults that are mostly beyond the distance, since you can assume that most of the quakes on those faults occurred outside the distance.

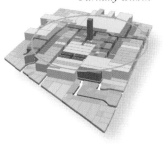

Partially within

If there is no direct interaction, you might want to include only features falling completely within the distance. For example, when analyzing road segments by the average number of accidents per month to see if accidents form clusters, you'd exclude road segments falling partially within the distance, since—unlike with earthquakes occurring on nearby faults—the fact that one segment has a lot of events doesn't influence another segment to have a lot of events.

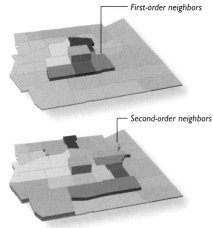

First-order neighbors

Second-order neighbors

If you're using a neighborhood based on adjacency, you can include only those features immediately adjacent to each other (termed first-order features), or you can include features that are once-removed, twice-removed, or more. The latter are termed higher-order features. Second-order features would be those adjacent to the first-order features. Using higher-order features allows you to increase the extent of adjacency-based neighborhoods.

If you do use higher-order features, you can specify whether or not to include lower-order features in the neighborhood as well. Including lower-order features allows you look for patterns at different scales. Excluding lower-order features allows you to look at the dispersion of a geographic phenomenon over time—you could see if the pattern changes as you include neighbors at higher orders. If, for example, you're trying to determine how quickly an outbreak of flu is spreading, you'd run the analysis using first-order census tracts, then run it again using only second-order tracts, and so on. In this case, you're using the increasing orders as a surrogate for the passage of time.

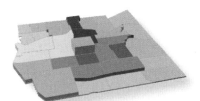

Neighborhood using first- and second-order features

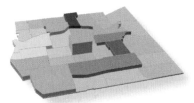

Neighborhood using second-order features only

If you're using an adjacency-based neighborhood, you need to decide if you will include only features that have a shared border, or if you will also include features that join at a corner. A shared border implies that the adjacent areas have more in common, or at least greater opportunity for the exchange of goods, people, ideas, and so on, than areas that simply meet at a corner. For example, when analyzing the vote in an election by precinct you'd probably include only features with shared borders in the neighborhood, since there is likely more opportunity for voters in these precincts to influence each other.

On the other hand, if the relationship is based on proximity rather than direct interaction between features, you'd include features that touch at corners, as well. For example, if you're analyzing the distribution of expensive land parcels to see if they tend to cluster, you'd include parcels that share a border or touch at corners when you define the neighborhood. The fact that the parcels are in proximity, rather than whether they touch at the corner or share a boundary, is the pertinent issue in establishing whether or not there are clusters.

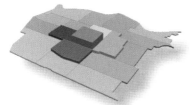
Shared borders only

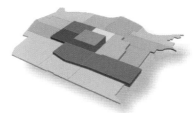
Shared borders and corners

The distinction is clear cut if you're analyzing rasters where every cell either has a fully shared border or is joined at a corner. If you're analyzing area features you'll likely find the distinction is less clear, with many features sharing a partial border. In this case, you can use as a weight value the percentage of the border that is shared. This is a more precise measure of the interaction between the areas than simply indicating whether or not any portion of the border is shared.

For adjacency, areas must be contiguous to be included as neighbors. If there is a gap between features, such as a street separating parcels, the weight for those neighbors will be 0. If you want the features to be included, use a distance-based neighborhood. In some cases, the feature separating the areas may represent a legitimate barrier that prevents features from being considered neighbors, such as a river that separates ZIP Codes.

ZIP Codes separated by a river

You can specify the rate at which the influence or interaction decreases as the distance increases. This is termed the distance decay.

If the influence decreases at a constant rate as the distance increases, the distance is the weight. This is termed linear distance decay. However, if you just used the measured distance as is, features that are farthest away would have the largest distances and the greatest weight, which is the opposite of what you want.

To give the closest features the greatest weight, the inverse of the distance is used (distance divided into one). Hence this method is referred to as "inverse distance" weight.

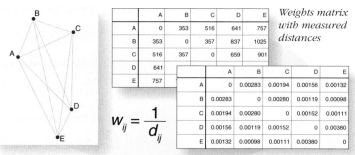

Weights matrix with measured distances

$$w_{ij} = \frac{1}{d_{ij}}$$

Weights matrix with inverse distances

If the influence decreases more (or less) rapidly as the distance increases, you assign an exponent to the distance based on the behavior of the geographic phenomenon being studied. This is termed exponential distance decay. For example, you could specify an exponential decay if you know that the value of a parcel is influenced by the values of all parcels in a city, but as the distance increases, the influence drops off more rapidly. If you can quantify the relationship between distance and the influence of features on each other, you should use this value for the exponent. In many cases, though, it's difficult to calibrate this. So, in practice, an exponent of 2 is often used. In this case, the distance would be squared before being divided into 1.

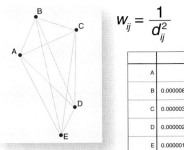

$$w_{ij} = \frac{1}{d_{ij}^2}$$

	A	B	C	D	E
A	0	0.0000080	0.0000038	0.0000024	0.0000017
B	0.0000080	0	0.0000078	0.0000014	0.0000010
C	0.0000038	0.0000078	0	0.0000023	0.0000012
D	0.0000024	0.0000014	0.0000023	0	0.0000145
E	0.0000017	0.0000010	0.0000012	0.0000145	0

Weights matrix with inverse squared distances

Alternatively, you can try several different values for the exponent to see what effect changing the value has on the results.

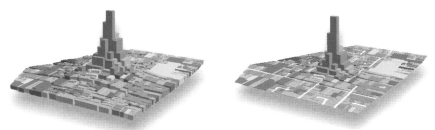

The height of each area represents its relative weight, with the tallest area being the target feature. With inverse squared distances (shown on the right), the influence of neighbors drops off more rapidly than with inverse distance weights (shown on the left), and distant features exert little influence on the target.

IS THE INFLUENCE DIVIDED PROPORTIONALLY AMONG THE NEIGHBORS?

For some geographic phenomena, the neighboring features each exert a fractional influence on the value of the target feature. Geographers Jay Lee and David Wong, in their book *Statistical Analysis with ArcView® GIS*, present the example of a house value being influenced in equal parts by the values of the surrounding houses. In these cases, you can specify a proportional weight. The main advantage of a proportional weight is that all target features get equal representation, no matter how many neighbors they have. Otherwise, target features having a higher number of neighbors will get greater representation when the values are summed.

For adjacency-based neighborhoods, each adjacent feature gets a weight based on the number of features adjacent to the target feature, rather than each adjacent feature getting a weight of 1. The sum of the weights of the adjacent features—each row in the matrix—always equals 1. For example, if a target area has four areas adjacent to it, each would get a weight of 0.25.

If another area has five adjacent areas, each would get a weight of 0.20. This method is termed "row-standardized" weighting.

Adjacency-based weights matrix, with total number of neighbors shown in the last column (top), and row-standardized weights matrix (bottom)

	1101	1201	1301	1102	1202	1302	1000	0901	0902	Row Sum
1101	0	1	0	1	1	0	0	0	0	3
1201	1	0	1	1	1	1	0	0	0	5
1301	0	1	0	0	1	1	0	0	0	3
1102	1									
1202	1									
1302	0									
1000	0									
0901	0									
0902	0									

	1101	1201	1301	1102	1202	1302	1000	0901	0902
1101	0.00	0.33	0.00	0.33	0.33	0.00	0.00	0.00	0.00
1201	0.20	0.00	0.20	0.20	0.20	0.20	0.00	0.00	0.00
1301	0.00	0.33	0.00	0.00	0.33	0.33	0.00	0.00	0.00
1102	0.25	0.25	0.00	0.00	0.25	0.00	0.25	0.00	0.00
1202	0.14	0.14	0.14	0.14	0.00	0.14	0.14	0.14	0.00
1302	0.00	0.25	0.25	0.00	0.25	0.00	0.00	0.25	0.00
1000	0.00	0.00	0.00	0.25	0.25	0.00	0.00	0.25	0.25
0901	0.00	0.00	0.00	0.00	0.25	0.25	0.25	0.00	0.25
0902	0.00	0.00	0.00	0.00	0.00	0.00	0.50	0.50	0.00

Row-standardized weighting is mostly used with adjacency-based neighborhoods to create proportional weights when features have unequal numbers of neighbors. You can also use row-standardized weights with distance-based neighborhoods. For distance-based neighborhoods, the weight (distance) to each feature is divided by the total distance for the row. In a distance-based neighborhood that covers the entire region, all features have an equal number of neighbors (that is, all other features in the region), so using row-standardized weights is not an advantage in terms of equalizing the influence of neighbors. However, for distance-based neighborhoods, row-standardized weights make the distances relative rather than absolute by scaling all the distances to 0 to 1. The effect of this is that near features are given a relatively greater weight, hence the influence drops off more rapidly as distance increases than if you used absolute weights.

According to some researchers, if the weight represents actual distance (as opposed to some other measure of influence, such as attraction), your results will be more accurate if you don't use row-standardized weighting with inverse distance. Otherwise, using row-standardized weights will give more accurate results.

When using an inverse distance weight with discrete features, the results can be distorted if there are features that are very close or at the same location. These features will have a very large weight compared to more distant features. That's because, as the points get closer and the distance between them approaches 0, the weight becomes very large. For example, if the distance between two points is 0.0001 meter—that is, they are essentially at the same location—the weight is 1/0.0001, which equals 10,000. By comparison, features at a distance of 100 meters would have a weight of 0.01 (1/100). So, one pair of points that is very close could distort the results.

ARE THERE DISCRETE FEATURES THAT ARE VERY CLOSE TO EACH OTHER OR IN THE SAME LOCATION?

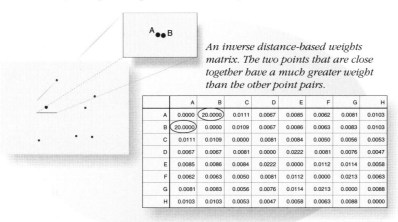

An inverse distance-based weights matrix. The two points that are close together have a much greater weight than the other point pairs.

	A	B	C	D	E	F	G	H
A	0.0000	20.0000	0.0111	0.0067	0.0085	0.0062	0.0081	0.0103
B	20.0000	0.0000	0.0109	0.0067	0.0086	0.0063	0.0083	0.0103
C	0.0111	0.0109	0.0000	0.0081	0.0084	0.0050	0.0056	0.0053
D	0.0067	0.0067	0.0081	0.0000	0.0222	0.0081	0.0076	0.0047
E	0.0085	0.0086	0.0084	0.0222	0.0000	0.0112	0.0114	0.0058
F	0.0062	0.0063	0.0050	0.0081	0.0112	0.0000	0.0213	0.0063
G	0.0081	0.0083	0.0056	0.0076	0.0114	0.0213	0.0000	0.0088
H	0.0103	0.0103	0.0053	0.0047	0.0058	0.0063	0.0088	0.0000

There is no ideal solution to this dilemma. One common fix is to add some value (usually 1) to all the distances before calculating the inverse. Another is to use a minimum threshold distance—any distance below the threshold is set to the minimum distance. Again, a value of 1 is often used.

In the former case, using the example above, the points that are 0.0001 meter apart would now have a distance of 1.0001 and an inverse distance (weight) of 0.99—a more reasonable weight than 10,000. The points that are 100 meters apart would now have a distance of 101, and a weight of 0.009, very close to the original weight value.

In the latter case, the points at a distance of 0.0001 would have a distance of 1 and a weight of 1; the points at a distance of 100 meters would maintain their original distance and weight (100 and 0.01, respectively).

References

Getis, Arthur, and Jared Aldstadt. "Constructing the Spatial Weights Matrix Using a Local Statistic." *Geographical Analysis* 36, no. 2 (2004): 90–104.

Lee, Jay, and David W. S. Wong. *Statistical Analysis with ArcView GIS*. Wiley, 2001.

4

Identifying clusters

Identifying clusters allows you to map
hot spots and cold spots. By comparing
the clusters to the locations of other
features you can better understand
why the clusters occur and decide
what action to take.

In this chapter:

• Why identify spatial clusters?
• Using statistics to identify clusters
• Finding clusters of features
• Finding clusters of similar values

Clusters occur in a geographic distribution when features are found in close proximity or when groups of features with similarly high or low values are found together (hot spots and cold spots). Identifying whether—and where—clusters exist is useful if you need to take action based on the location of one or more clusters—such as assigning a task force to deal with a cluster of burglaries.

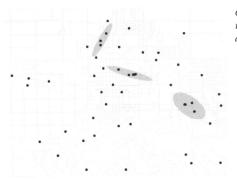

Clusters of burglaries identified using statistical analysis

Pinpointing the locations of clusters can help you examine the causes of the clustering. By comparing the locations of clusters to the other features, you can start to identify possible contributing factors. For example, by comparing the clusters of cases of a particular disease to environmental and economic data, you could see if there are possible spatial relationships between the clusters and any of these factors.

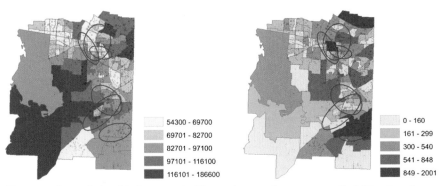

54300 - 69700	0 - 160
69701 - 82700	161 - 299
82701 - 97100	300 - 540
97101 - 116100	541 - 848
116101 - 186600	849 - 2001

There may be a relationship between significant clusters of emergency calls (ellipses) and median home value (shown by block group, left), but not necessarily with numbers of young adults (right).

By looking at a map, you can draw conclusions about where there are clusters of features. Statistics lets you test those conclusions and validate them by measuring whether features are closer than would occur by chance. You can then map the results of the test, not just the features themselves.

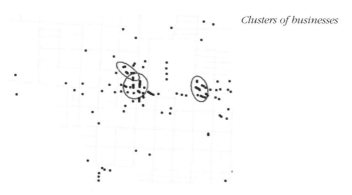

Clusters of businesses

Using statistics takes much of the guesswork out of identifying clusters. If there are multiple events at a single location, such as several auto thefts at the same address, the clusters can be hard to see if you simply map the features. When you use statistics to identify clusters, each event is counted as a unique occurrence.

Several thefts at one address

Locations of auto thefts (left) and clusters of auto thefts (right)

When using statistics to identify clusters, you can calculate the probability that the clusters are not due to chance, so you can be more confident in any decisions you make based on the results of the analysis.

Clusters can be formed using the locations of features alone, or formed using the location influenced by an attribute value (clusters of features with similar values).

Clusters based on location alone are composed of discrete features. Often the features are points, such as auto accidents, earthquake epicenters, or cases of measles.

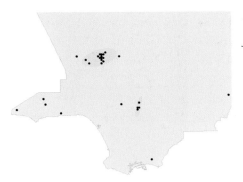

Earthquakes in Los Angeles County in March 1974, forming two clusters

The features can also be discrete areas, such as cut areas in a forest. However, since the centroids of the areas are used to identify the clusters, the results may be misleading, especially if the areas are elongated or convoluted. (See "Using statistics with geographic data.")

Clusters of features having similar attribute values can be composed of discrete features (points or areas), spatially continuous data, or data summarized by contiguous areas (such as census tracts or counties). The attributes are interval or ratio values.

A market analyst might locate clusters of coffee shops with low sales, to find out where additional marketing is needed. A social service agency planning senior services could identify clusters of census tracts where a high percentage of the population is seniors.

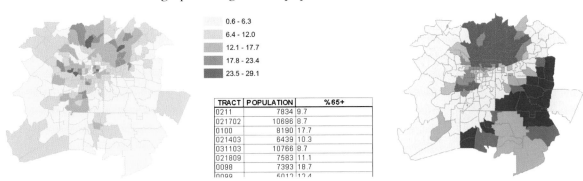

	0.6 - 6.3
	6.4 - 12.0
	12.1 - 17.7
	17.8 - 23.4
	23.5 - 29.1

TRACT	POPULATION	%65+
0211	7834	9.7
021702	10696	8.7
0100	8190	17.7
021403	6439	10.3
031103	10766	8.7
021809	7583	11.1
0098	7393	18.7
0099	5012	12.4

Percent age 65 and over by census tract (left) and statistically significant clusters of tracts with a high percentage of seniors (orange) or a low percentage (blue).

What's the time period of the data?

By identifying a cluster, in many cases you're assuming the features are related in time as well as in space. For static features, such as vacant parcels, you use a snapshot of the current condition. If you're analyzing events that individually take up little time—such as crimes or earthquakes—you need to define the period to use. For example, a crime analyst looking for clusters of gang-related assaults might include incidents for the past five or six months; assaults from longer ago are unlikely to be related to any current incidents. On the other hand, a geologist looking for clusters of earthquakes over time, to possibly predict future seismic activity, would likely include events for the entire available period of record—at least a hundred years.

Even for the same type of feature, the time period will change depending on the purpose of your analysis. A crime analyst trying to find burglaries possibly committed by the same group of people would include only recent burglaries. If the analyst were trying to identify clusters representing ongoing high-crime areas, she'd include burglaries over several years in the analysis.

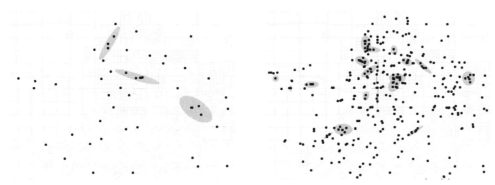

Cluster of burglaries occurring over one month (left) and over one year (right).

What measure of distance are you using?

Clusters are usually defined using straight-line (Euclidean) distance between features. However, you can also use other measures of distance, such as travel time or cost. For example, a crime analyst would want to identify clusters of burglaries using driving time between the crimes rather than straight-line distance, especially if the incidents are separated by a barrier. Two burglaries on opposite sides of the river may be near each other using straight-line distance but in fact may not be very close in terms of travel time.

One method for finding clusters of discrete features is to specify the distance features can be from each other in order to be part of a cluster, and the minimum number of features that make up a cluster. The method, described by CrimeStat author Ned Levine, is known as nearest neighbor hierarchical clustering.

Moose sightings during March, over several years (left) and clusters of sightings (right)

Nearest neighbor hierarchical clustering uses the distance between features (similar to nearest neighbor analysis for analyzing the pattern of a distribution of features).

Clusters of businesses. Features must be within a specified distance of each other to be considered part of a cluster.

It is hierarchical because, after delineating a first set of clusters, the routine continues on to group the clusters into larger clusters.

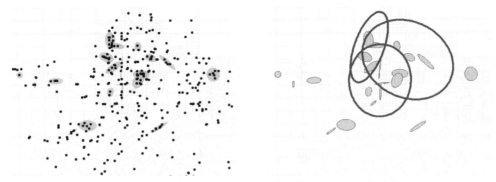

Clusters of burglaries (left) and the clusters grouped into three larger clusters (right)

Hierarchical clustering shows you how the features are clustered at several geographic scales, such as how burglaries cluster at both the neighborhood and citywide scale. In crime analysis, for example, as Ned Levine describes in his CrimeStat documentation, you could identify first-order clusters at the neighborhood level, where officers would intervene for specific incidents, and higher-order clusters, which might correspond to areas requiring community policing.

In wildlife management, clusters of individuals of a species might identify habitat areas that need to be protected, while clusters of these habitat areas might define larger management areas requiring development of corridors linking the clusters, and other strategies.

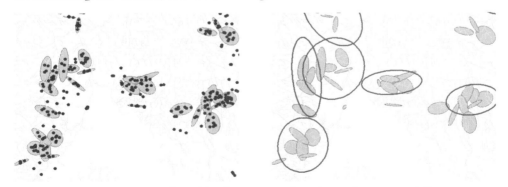

Clusters of wolverine sightings (left) and the clusters grouped into larger clusters, potentially defining habitat areas (right)

HOW NEAREST NEIGHBOR HIERARCHICAL CLUSTERING WORKS

For nearest neighbor hierarchical clustering, you specify a probability level, which the GIS uses to calculate the distance within which features will be considered a cluster. With a lower probability, more features will be included in a cluster, but you'll be less certain that the features actually represent a cluster.

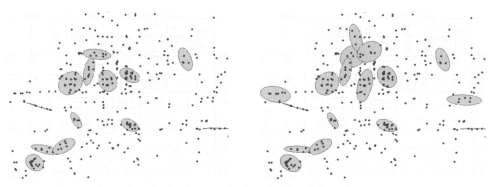

Clusters of assaults calculated using a high probability (left) and a low probability (right)

There is a range within which the distance between two features may occur owing simply to chance. This range—termed the confidence interval—is calculated based on the probability level you specify.

If the distance is greater than the high end of the range, the features are farther apart than you would expect by chance. Since you're trying to find features that are closer than you would expect by chance (clusters), you're interested in the lower end of the range. This value is the minimum, or "threshold," distance.

The confidence interval is calculated using the mean distance that would occur between points in a random distribution—that is, the mean distance between points if there were the same number of points as in the dataset you're analyzing spread over the same study area, and you knew the points were randomly distributed. This is termed the "mean random distance."

The mean random distance is calculated by multiplying 0.5 by the square root of the area divided by the number of points. (This is the same calculation used for the nearest neighbor index described in chapter 3, "Measuring geographic patterns.")

The mean
random distance

$$d = 0.5 \sqrt{\dfrac{A}{n}}$$

The extent of the study area (A) is divided by the number of features (n), and the square root is taken....

....then the result is multiplied by 0.5

$$d = 0.5 \sqrt{\dfrac{139392000}{200}} = 417.4$$

Dividing the area by the number of points gives the average number of points per unit area for the study area. The square root is then taken to turn this into linear distance (from square meters to meters, for example). The value is cut in half since the calculation assumes distances are measured twice (from point A to point B and back).

The GIS then finds the confidence interval by calculating the standard error for the distribution of points. The standard error measures how much the mean random distance varies around its average.

0.26136 (a constant derived from the standard error for a normal curve) is divided by....

The Standard Error

$$SE = \dfrac{0.26136}{\sqrt{\dfrac{n^2}{A}}}$$

....the square root of the number of features (n) squared, divided by the extent of the study area (A)

$$SE = \dfrac{0.26136}{\sqrt{\dfrac{40000}{139392000}}} = 15.4$$

The GIS uses the probability level you specified to look up a value, known as t, in a standard table (the Student's t-distribution). The t-distribution is appropriate when the standard error for the mean of the dataset as a whole (the population) is an estimate, as is the case here. The specific value of t is based on the number of features in the dataset and the degrees of freedom. In this case, the degrees of freedom are assumed to be at least 120, which is appropriate for a relatively large sample (see the discussion of the Chi-square test in chapter 3, and the references at the end of that chapter, for more on degrees of freedom).

The t-value is multiplied by the standard error. The confidence interval is the mean random distance plus or minus this product. The threshold distance is the lower limit of this range.

For p = 0.05, t = 1.96

*t * SE = 1.96 * 15.4 = 30.2*

Confidence interval

387.2   d = 417.4   447.6

Threshold distance

The higher the probability, the wider the confidence interval and the lower the threshold distance—features have to be closer to be part of a cluster. The lower the probability, the narrower the confidence interval and the higher the threshold distance—features can be farther apart and still be included in a cluster.

If, for example, you want to be 95% sure the features in a cluster are not near each other simply by chance, you'd specify a probability of 0.05. If you want to be 80% sure, you specify a probability of 0.2. The range around the mean random distance is narrower, and the threshold distance is higher, including more features in the cluster. While there are more features in the cluster, you're less certain that they aren't there simply by chance.

For p = 0.20, t = 1.28

*t * SE = 1.28 * 15.4 = 19.7*

Confidence interval

397.7   d = 417.4   437.1

Threshold distance

Once the GIS calculates the threshold value, it measures the distance between each pair of points and assigns to a cluster all point pairs having a distance less than the threshold. So, as long as a point is within the threshold distance of at least one other point in a cluster, it will be included in that cluster. Any points not within the threshold distance of another point remain outside a cluster. At this stage, a cluster can consist of as few as two points.

You specify a minimum number of features required to form a cluster, based on your knowledge of the features you're analyzing. For example—depending on the crime rate in the area—two or three burglaries in a neighborhood over a several-month period may be too few incidents to be considered a hot spot, whereas five or six burglaries over the same period would represent a meaningful cluster.

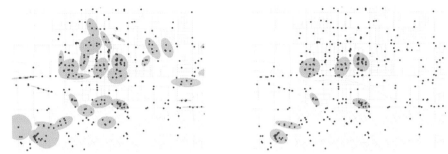

Clusters of assaults calculated using a minimum of four assaults per cluster (left) and a minimum of ten assaults per cluster. With a higher minimum, fewer clusters are created.

If a cluster has fewer than the minimum number of points you speci-fied, the cluster is dissolved. Since the points were not close enough to any other points to be included in another cluster, they remain outside a cluster.

Minimum points = 3

After creating the clusters, the GIS finds the median center for each (see chapter 2, "Measuring geographic distributions"), calculates a new con-fidence interval and threshold distance based on the centers, and then assigns the centers to new clusters. It continues the process, creating pro-gressively higher-order clusters, until all lower-order clusters have been grouped into a single cluster, or until it can no longer group cluster cen-ters into any new clusters, because the centers are beyond the threshold distance.

DISPLAYING THE CLUSTERS

Once points have been assigned to a cluster, individual clusters can be displayed. For example, the software might calculate the standard devi-ational ellipse for the points in each cluster. That makes it easy to see the orientation of individual clusters and of the clusters as a whole. The ellipses for higher-order clusters may show different trends at different scales.

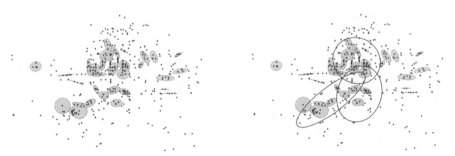

Clusters of assaults drawn using the standard deviational ellipse for each cluster. The map on the right shows the ellipses for the first- and second-order clusters.

If you display the clusters using standard deviational ellipses, you can specify how many standard deviations to use for the ellipse created for each cluster. One standard deviation will cover the features that represent the heart of the cluster and are more likely to be within the threshold distance of more than one other feature. Using two or more standard deviations will include features on the periphery of the cluster.

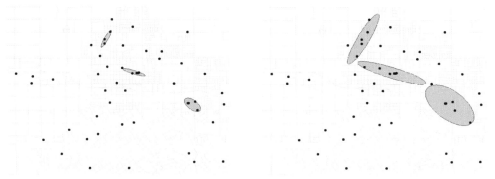

The ellipses for clusters of burglaries drawn using one standard deviation (left) and three standard deviations (right).

You could draw the ellipses for several standard deviations on a single map to get a better sense of the extent of each cluster and the distribution of features within each.

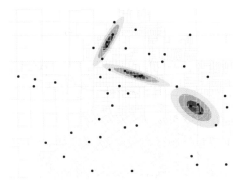

Clusters of burglaries drawn using ellipses calculated from one, two, and three standard deviations.

The clusters at a particular order might provide the most useful information. First-order clusters might be too small and scattered, while higher-order clusters too large to be useful in pinpointing hot spots. Using clusters between these extremes would provide the best information.

The many first-order clusters of assaults (left) overlap, making it difficult to spot the overall pattern. The third-order cluster (center) is too generalized to be useful. The second-order clusters (right) identify three major areas with high numbers of assaults.

Displaying the clusters lets you see how concentrated or dispersed they are. You may see, for example, "clusters of clusters."

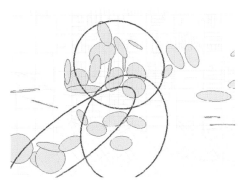

In this map of first- and second-order clusters of assaults, both individual hot spots and the overall pattern can be seen.

COMPARING THE CLUSTERS TO A CONTROL GROUP

If you're looking for causes of the clusters, you'll want to compare the clusters to a control group. For example, when analyzing assaults, you'd want to compare the clusters of assaults to where people live—there may be more assaults in an area simply because there are more people there. If there are more assaults than you'd expect based on the population density, there may be other factors at play. You could also use all crimes as the control to see if clusters of assaults occur outside of generally high-crime areas.

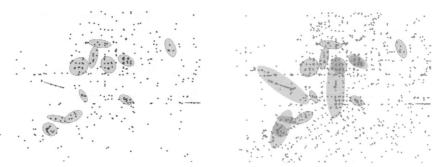

Clusters of assaults (left) and the clusters of assaults with clusters of all crimes, shown by red ellipses (right). Some assault clusters occur outside high-crime areas.

To compare the clusters to a control group, you can:

• Map the clusters and the control group together.

• Create clusters for the control group and compare these to the clusters of the features you're analyzing. If the control group data is summarized by area—for example, population density by block group—you can use one of the methods for finding clusters of similar values (described later in this chapter). The Gi* method works best for this since it shows clusters of high and low values.

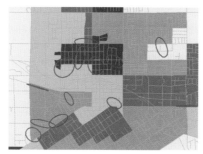

Clusters of assaults with population per square mile by block group. While most of the clusters occur where there is high population density, as you might expect, a few occur in areas of low population density, pointing to other causes.

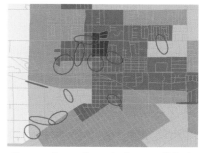

Clusters of assaults with clusters of block groups with high population density (dark color) or low (light color). The clusters of assaults that are in areas of low population density stand out.

- Summarize the features by the control group. For example, if you wanted to compare clusters of crimes to the distribution of people, you'd calculate the number of crimes per person for each census tract, and then find the clusters of tracts with high values for this attribute. Mapping clusters of features with high values is discussed later in this chapter.

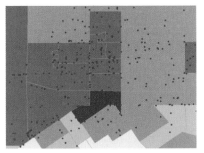

Assaults per person, by block group

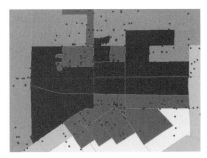

A cluster of block groups having a high number of assaults per person.

FACTORS INFLUENCING THE RESULTS OF NEAREST NEIGHBOR HIERARCHICAL CLUSTERING

The more features in an area, the smaller the threshold distance will be. That's because the mean random distance—which in turn determines the threshold distance for defining a cluster—is calculated using the number of features and the extent of the study area.

Assaults in two different portions of a city having approximately the same areal extent. The area on the left has fewer assaults, so the threshold distance is less, resulting in a cluster of assaults that are relatively far apart.

Since you specify the minimum distance between features and the minimum number of features required to create a cluster, the results are subjective. By changing the criteria you can alter the results.

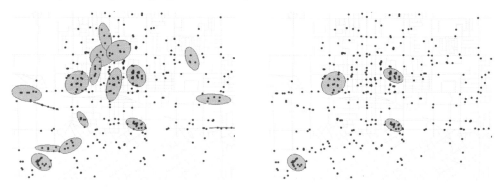

Clusters of assaults. Using a smaller distance between features and a higher minimum number of features per cluster results in fewer clusters (right).

If you're analyzing discrete areas, variation in the size of the features can affect the results. Nearest neighbor hierarchical clustering works best when the areas are of roughly equal size.

To find clusters of features that have similar values for a particular attribute, the GIS looks at the attribute values of each feature and its neighbors, as well as the proximity of the features.

The GIS calculates a statistic for each feature indicating the degree to which nearby features have similar values for a given attribute. You can then map these statistical values to see where there are clusters of features with similar values.

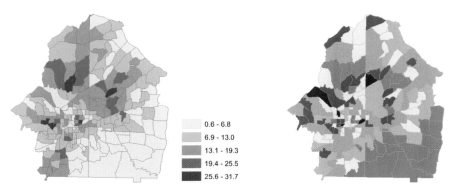

	0.6 - 6.8
	6.9 - 13.0
	13.1 - 19.3
	19.4 - 25.5
	25.6 - 31.7

Percent age 65 and over, by census tract (left), and tracts having a percentage of seniors similar to their neighbors (orange) or unlike their neighbors (blue). The darker the orange, the more similar the values; the darker the blue, the less similar.

Mapping features based on how similar they are to surrounding features is different than simply mapping the values of features. While mapping the percentage of people age 65 and over by census tract can show you where a high proportion of seniors live, you can use statistics to assign each tract a value quantifying how similar it is to its neighbors. By mapping the tracts using this new value, it's clear where clusters of tracts having particularly high or low percentages of seniors are. The clusters could define potential market areas for products targeted to seniors.

If you simply map the attribute values, the way you classify values will change the way features appear to be clustered or not. Using statistics gives you confidence that what looks like a cluster really is a cluster.

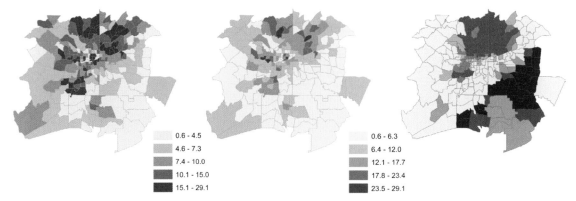

0.6 - 4.5	0.6 - 6.3
4.6 - 7.3	6.4 - 12.0
7.4 - 10.0	12.1 - 17.7
10.1 - 15.0	17.8 - 23.4
15.1 - 29.1	23.5 - 29.1

Census tracts color coded by percent age 65 and over. The map on the left, using a quantile classification, shows what appear to be several strong clusters of high values. In the middle map, which uses an equal interval classification, the clusters are less pronounced. The map on the right shows significant clusters of high (orange) and low (blue) values calculated from the data.

To determine the likelihood that a feature having values similar to its neighbors is not due to chance, you use a statistical test. You can then map the results of the test to see which clusters are statistically significant.

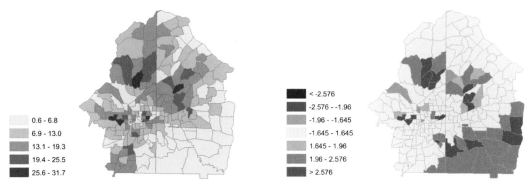

0.6 - 6.8	< -2.576
6.9 - 13.0	-2.576 - -1.96
13.1 - 19.3	-1.96 - -1.645
19.4 - 25.5	-1.645 - 1.645
25.6 - 31.7	1.645 - 1.96
	1.96 - 2.576
	> 2.576

Percent age 65 and over, by census tract (left), and tracts with values similar to their neighbors (orange) or unlike their neighbors (blue) at several critical levels of significance (right)

METHODS OF IDENTIFYING CLUSTERS OF SIMILAR VALUES

You can find out how much each feature is similar or dissimilar to its neighbors—where high values are surrounded by high values or low values are surrounded by low values. This method is useful for showing where there are areas with similar values and where similar and dissimilar values are interspersed.

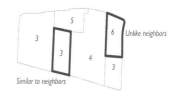

The method calculates a statistic for each feature. You can then map the features based on this value to see the locations of features with similar values. To see whether the clusters represent high values or low values, you can map the values and display this map along with the map of the statistic.

You can also find out where hot spots and cold spots are. This method looks at values of adjacent features or features within a distance you specify and compares the average value for the neighborhood to the average value for the study area. It indicates whether clusters are composed of high or low values.

	0.0 - 0.9
	1.0 - 2.4
	2.5 - 4.3
	4.4 - 6.8
	6.9 - 13.2

Percent Asian population by census tract (left map). One method (middle) shows tracts that have values similar to, or unlike, their neighbors (dark blue and light blue, respectively). Another method (right) shows tracts with similarly high (dark) or low (light) density of Asian population.

These methods are adaptations of the global methods, discussed in chapter 3, for analyzing patterns and are termed local statistics (since they show local variation across the study area). The global methods calculate a single statistic that summarizes the pattern for the study area, while the local methods calculate a statistic for each feature based on its similarity to its neighbors.

The local methods were developed primarily by economist and geographer Luc Anselin and by regional scientists Art Getis and Keith Ord in the mid-1990s. The purpose was—in addition to identifying clusters—to look at how individual features contribute to the pattern of values identified by the global methods, and to look for spatial anomalies or outliers, where one or a few features may have values very different from nearby features. This use of the statistics is especially useful to test for spatial autocorrelation before running a statistical model. That is, if a global statistic indicates there is spatial autocorrelation, the measure of local variation can help pinpoint which feature or features are contributing to it, so you can account for it in the model. (See chapter 5, "Analyzing geographic relationships," for more on statistical models.)

IDENTIFYING SIMILAR VALUES AMONG NEIGHBORING FEATURES

The two statistics for identifying similar values, Geary's c and Moran's I, have local versions. Local Geary's c (or Geary's c_i) compares the values of neighboring features by calculating the difference between them. (The i subscript indicates that a statistic is calculated for each feature.) Moran's I_i compares each value in the pair to the mean value for all the features in the study area.

Geary's c_i emphasizes how features differ from their immediate neighbors, since it compares the values of neighboring features directly to each other, whereas Moran's I_i emphasizes how features differ from the values in the study area as a whole, since it compares the value of each feature in a pair to the mean value for all features in the study area.

Some researchers believe that the results obtained from Moran's I_i are more useful than those from Geary's c_i, and also that the significance test for Moran's I_i is more reliable. While this chapter focuses on Moran's I_i, several of the references listed at the end of the chapter contain information about Geary's c_i.

Defining the neighborhood for Moran's I_i

Since, when using Moran's I_i, you're interested in local variation (what's happening right around each feature), a neighborhood based on adjacent features is appropriate. In contrast, a neighborhood based on distance takes into account the influence of all the features in the dataset, some of which may be very distant.

If you use a neighborhood based on adjacency, you can use either binary weighting or row-standardized weighting. With row-standardized weighting, each neighbor gets an equal weight based on the number of neighbors surrounding the target feature. With binary weighting, features with more neighbors will have a higher total value. (Each neighbor gets a weight of 1.) Because of this, features near the edge of the study area—which will have fewer neighbors—are likely to have lower values for I_i. Using row-standardized weights avoids this problem. (See "Defining spatial neighborhoods and weights.")

Calculating Moran's I_i

For Moran's I_i, the values of the target feature and the neighboring features are both compared to the mean.

The GIS first calculates the mean value for the attribute you're analyzing. Then it calculates the difference from the mean for each neighbor and multiplies it by the weight for that neighbor. Next it sums these products. Finally, it multiplies the sum by the ratio of the difference from the mean for the original feature's attribute value, divided by the variance (see "Understanding data distributions").

Local Moran's I, calculated for each feature (i)

The mean value (x) is subtracted from the value of the neighbor (x) and the difference multiplied by the weight (w) for the target-neighbor pair; the results for all neighbors are summed....

$$I_i = \frac{(x_i - \bar{x})}{s^2} \cdot \sum_j w_{ij} (x_j - \bar{x})$$

....then the sum is multiplied by: the difference between the mean value (x) and the target feature value (x), divided by the variance (s²)

The ratio of the difference from the mean divided by the variance is a constant (that is, the same value is used for each calculation of I_i). The constant is used to scale each I_i value to the value of the global I statistic. Otherwise, you'd have a huge range of values for I_i. Essentially, Local Moran's I represents a disaggregation of the global version of Moran's I into its component parts.

The equation is often shown substituting z_i for $(x_i - \bar{x})$ and z_j for $(x_j - \bar{x})$. By calling both of them z, it's clear that it's the same function, just used for different features (i and j, respectively).

The standard form of Local Moran's I; z represents the difference in value between the target and the mean, while z represents the difference in value between each neighbor and the mean

$$I_i = (z_i / s^2) \sum_j w_{ij} z_j$$

Interpreting Moran's I_i

A large positive value for Moran's I_i indicates that the feature is surrounded by features with similar values, either high or low.

A negative value for I_i indicates that the feature is surrounded by features with dissimilar values.

Several adjacent features with high values for I_i defines a cluster of similar values. The statistic doesn't indicate if the attribute values themselves are high or low.

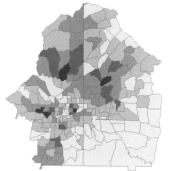

	0.6 - 6.8
	6.9 - 13.0
	13.1 - 19.3
	19.4 - 25.5
	25.6 - 31.7

TRACT	% 65+	LOCAL I
010103	19.08	0.119
021205	5.59	-0.153
021211	7.31	0.448
021210	8.03	0.173
010204	8.18	-0.192
010205	16.43	0.282
021207	13.94	-0.015
021304	4.20	0.600

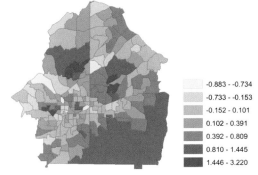

	-0.883 - -0.734
	-0.733 - -0.153
	-0.152 - 0.101
	0.102 - 0.391
	0.392 - 0.809
	0.810 - 1.445
	1.446 - 3.220

Percent age 65 and above, by census tract (left), and tracts color coded by I_i values (right). Dark orange indicates tracts with high values for I_i, indicating they are surrounded by tracts having a similar percentage of seniors.

The magnitude of the I_i value (either high or low) depends on the difference in attribute values, the number of neighbors with similar values, and the magnitude of the attribute values.

- Neighboring values that are very close or even the same as the target feature will result in a higher I_i value for the feature; similarly, the larger the difference in dissimilar values, the lower the I_i value will be.

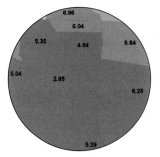

Tracts with a similar percentage of senior citizens (labels) all have high I_i values (dark orange).

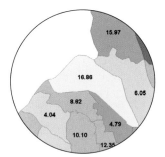

The tract with a high percentage of seniors (16.86%) gets a low relative I_i value (pale orange) since it's surrounded by tracts with lower percentages.

- One neighbor with a very dissimilar value could result in the feature having a low I_i, even if the rest of the neighbors have relatively similar values.

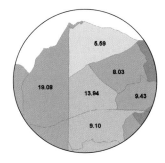

The tracts shown in light orange have low I_i values— they're adjacent to a tract with a much higher percentage of seniors (19.08%).

- For similar values, higher highs and lower lows will result in a larger value for I_i.

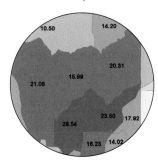

The two tracts with a very high percentage of seniors (28.54% and 23.50%)—and surrounded by tracts that also have high percentages—get very high I_i values (shown in dark orange).

Once you've calculated the I_i statistics for the features, you can map each feature based on the value of the statistic. You can then see where the clusters of features with similar values are, or where there may be neighboring features with very dissimilar values. If you're analyzing area features, you can simply create a map by classifying the I_i values into ranges.

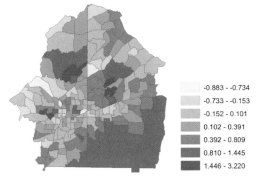

	-0.883 - -0.734
	-0.733 - -0.153
	-0.152 - 0.101
	0.102 - 0.391
	0.392 - 0.809
	0.810 - 1.445
	1.446 - 3.220

Census tracts color coded by I_i values calculated from percent age 65 and above

Testing the statistical significance of Moran's I_i

As with the global Moran's I statistic, you can measure whether each value of I_i is statistically significant at a given confidence level. What you're measuring is the likelihood that the similarity between a feature and its neighbors isn't due simply to chance. You do this by calculating a Z-score, which tells you the likelihood you'd be wrong to reject the null hypothesis. The null hypothesis for the test depends on the sampling assumption you use—either normalization or randomization. (See "Testing statistical significance.")

First, the expected I_i—assuming a random distribution of values—is calculated. This is then subtracted from the observed I_i, and the difference divided by the square root of the variance (the average amount the values vary from the mean).

The Z-score for Moran's I_i

For each feature, the expected I_i value is subtracted from the observed I_i....

$$Z(I_i) = \frac{I_i - E(I_i)}{\sqrt{Var(I_i)}}$$

....and the difference divided by the square root of the variance

The expected I_i value is calculated by taking the negative of the sum of the weights and dividing by the number of features, minus 1.

The expected Local I value

The negative of the sum of the weights for all feature pairs (w_{ij})...

$$E(I_i) = \frac{-\sum_j w_{ij}}{n-1}$$

....is divided by the number of features (n) minus one

When the GIS calculates the Z-score, it subtracts the expected value from the observed value. The expected value incorporates the number of neighbors (represented by the sum of the weights)—it is the value of I_i you would expect, given the number of neighbors for that feature and the variance. For binary weighting, the sum of the weights is the number of features (since each neighbor has a weight of 1); for row-standardized weighting, the sum is always 1. So, the Z-score is essentially scaled by the number of neighbors. If you have a feature with only one or two neighbors, you'd expect the I_i value to be small when using binary weighting. If the I_i value is large, and the variance is small, the clustering is likely significant. On the other hand, if the I_i value and the variance are both large, then the large I_i value is probably less significant.

A high positive Z-score for a feature indicates the surrounding values are similar values, high or low. So, a group of adjacent features having high Z-scores indicates a cluster of similarly high or low values. A very negative Z-score for a feature indicates the feature is surrounded by dissimilar values—a high value surrounded by low values, or vice versa. See "Testing statistical significance" for more on Z-scores.

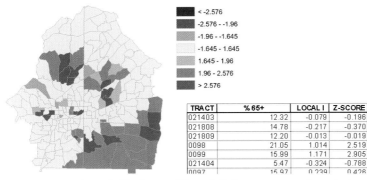

	< -2.576
	-2.576 - -1.96
	-1.96 - -1.645
	-1.645 - 1.645
	1.645 - 1.96
	1.96 - 2.576
	> 2.576

TRACT	% 65+	LOCAL I	Z-SCORE
021403	12.32	-0.079	-0.196
021808	14.78	-0.217	-0.370
021809	12.20	-0.013	-0.019
0098	21.05	1.014	2.519
0099	15.99	1.171	2.905
021404	5.47	-0.324	-0.788
0097	15.97	0.239	0.426

Census tracts color coded by Z-score (Moran's I_i) for percentage of population age 65 and over

Mapping the I_i values shows you where clusters of similar values are; mapping the Z-scores shows which clusters are statistically significant. And mapping the attribute values themselves shows you whether the cluster is comprised of high or low values. So you really need to map all three values side by side. The software adds the I_i statistic and the Z-score to each feature's record in the layer's attribute table.

	0.6 - 6.8
	6.9 - 13.0
	13.1 - 19.3
	19.4 - 25.5
	25.6 - 31.7

	-0.883 - -0.734
	-0.733 - -0.153
	-0.152 - 0.101
	0.102 - 0.391
	0.392 - 0.809
	0.810 - 1.445
	1.446 - 3.220

	< -1.645
	-1.645 - 1.645
	> 1.645

Census tracts showing percent age 65 and above (left), I_i values for tracts (middle), and Z-scores for tracts (right), showing which clusters are statistically significant at a confidence level of 0.10.

If you're using row-standardized weights, you could reclassify the I_i values as being above or below the expected value for a random distribution. (The expected value is the same for all features when using row standardization.) That will show whether the observed values are higher or lower than expected for a random distribution. (See "Defining spatial neighborhoods and weights" for more on row-standardized weighting.)

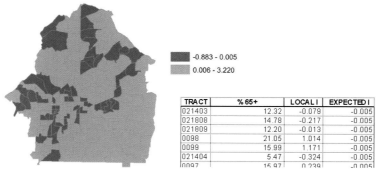

TRACT	% 65+	LOCAL I	EXPECTED I
021403	12.32	-0.079	-0.005
021808	14.78	-0.217	-0.005
021809	12.20	-0.013	-0.005
0098	21.05	1.014	-0.005
0099	15.99	1.171	-0.005
021404	5.47	-0.324	-0.005
0097	15.97	0.239	-0.005

Color legend:
- -0.883 - 0.005
- 0.006 - 3.220

Census tracts color coded to show I_i values above the expected I_i for a random distribution (orange) and below it (blue), for population age 65 and over

You can then classify the Z-scores as being less than, within, or greater than the confidence interval for a given confidence level to show which I_i values are significant.

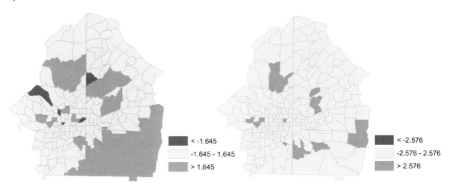

Left map legend:
- < -1.645
- -1.645 - 1.645
- > 1.645

Right map legend:
- < -2.576
- -2.576 - 2.576
- > 2.576

Census tracts color coded to show Z-scores above the critical values (orange) and below them (blue) at a confidence level of 0.10 (left map) and 0.01 (right map). To see which clusters are high values and which are low, you need to compare this map to the map of percentage of population age 65 and over.

To see the clusters at several confidence levels—say 90%, 95%, and 99%—on a single map, you'd classify the Z-scores using several ranges. That would show you the clusters and indicate which are most significant.

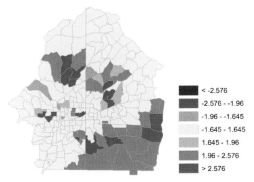

■	< -2.576
■	-2.576 - -1.96
■	-1.96 - -1.645
□	-1.645 - 1.645
■	1.645 - 1.96
■	1.96 - 2.576
■	> 2.576

Census tracts showing Z-scores classified by critical values for several confidence levels, calculated from percent age 65 and above

To see—on a single map—where the clusters are, which ones are significant, and whether they're clusters of high or low values, you'd map multiple variables. To do this, you may need to reclassify the I_i values and the Z-scores into two or three classes at most. Otherwise, the map will have too many colors to be readable.

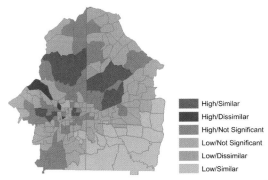

■	High/Similar
■	High/Dissimilar
■	High/Not Significant
■	Low/Not Significant
■	Low/Dissimilar
□	Low/Similar

Dark orange indicates a tract with a high percentage of seniors that is significantly similar to its neighbors (at a confidence level of 0.10); light orange indicates similar tracts having a low percentage of seniors. Blue tracts are unlike their neighbors, either because they have a higher (dark blue) or lower (light blue) percentage of seniors.

Factors influencing the results of Moran's I_i

As with other spatial statistics, the results near the edges of the study area may be suspect. In the case of Moran's I_i, the features near the edge have fewer neighbors than features in the interior, so differences or similarities among surrounding features may be exaggerated. (See "Using statistics with geographic data.")

Features surrounded by neighbors having a range of high and low values may receive a high value for I_i, indicating surrounding features have similar values. This is particularly true if the high values are closer to the feature's value than the low values are. You'll want to look closely at the clusters to see what the underlying attribute values are before making any decision based on the analysis.

When you're analyzing a small number of features—fewer than 30—the results will be suspect, since any outliers are likely to skew the distribution of values. (See "Understanding data distributions.")

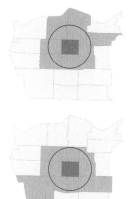

The Z-score can produce misleading results when used with local statistics, such as Moran's I_i. That's because the test assumes independence between features. But because the GIS runs the test to calculate a Z-score for each feature, the test will end up using many of the same neighbors for adjacent features.

That violates the independence of the test, especially since the values of features that are adjacent are likely to be more similar in any case. Researchers Art Getis and Keith Ord have suggested an alternate test, called a Bonferroni-type test, to overcome this issue. This test uses more stringent criteria to ensure that the results are significant at the confidence level you specify.

Adjacent features share some neighbors.

By changing the critical value, the Bonferroni correction (named for the Italian mathematician who developed the method in the 1930s) makes it more difficult for any one test to be statistically significant. In the correction's simplest form, you divide the original confidence level by the number of tests to get an adjusted confidence level. With a confidence level of 0.10 and 30 features in the dataset, the adjusted confidence level would be 0.003 (0.10/30). You'd compare the calculated Z-score for each feature to the critical value obtained using this adjusted confidence level.

IDENTIFYING CONCENTRATIONS OF HIGH OR LOW VALUES WITHIN A DISTANCE

This method, termed the local G-statistic, shows you where clusters of high values or low values are. For each feature, the statistic compares neighboring features within a distance that you specify. The statistic indicates the extent to which each feature is surrounded by similarly high or low values. A value is generated for each feature, or each cell in a raster surface. You can then map the features color coded by these values to see the clusters.

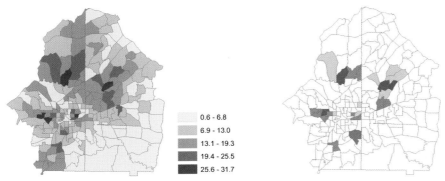

	0.6 - 6.8
	6.9 - 13.0
	13.1 - 19.3
	19.4 - 25.5
	25.6 - 31.7

Percent age 65 and over, by census tract (left). The map on the right shows clusters of tracts having a high percentage of seniors (orange) and tracts having a significantly lower percentage than their neighbors (blue).

Because you specify a distance within which to look for similar values, you can measure the similarity for a range of distances. That lets you identify the distance at which the clustering is greatest.

Two versions of the local G-statistic

There are two versions of this statistic, both developed by Art Getis and Keith Ord. In one version, the value of the target feature itself is not included in the equation. This is the Gi statistic. You'd use the Gi statistic if you're interested in the effect of the target feature on what's going on around it. This would be the case if you're interested in the dispersion of a particular phenomenon from the target feature to the surrounding area over time. Getis and Ord, for example, used Gi to track the dispersion of AIDS to counties surrounding San Francisco County over the course of several years. They wanted to see if the intensity of clustering of AIDS cases in counties surrounding San Francisco increased over time and the distance at which the clustering peaked. See the references at the end of this chapter for more on the Gi statistic.

In the other version, called Gi* (pronounced G-i-star), the value of the target feature is included. If you're interested in finding hot spots or cold spots, you'd use Gi*—you'll want to include the value of the target feature since its value contributes to the occurrence of the cluster.

Defining the neighborhood for Gi*

Gi* uses a neighborhood based either on adjacent features or on a set distance. When using a distance-based neighborhood, the distance you specify is based on your knowledge of the features and their behavior. For example, if you're looking for locations to build pet stores and you have survey data showing ZIP Codes where there are high numbers of pet owners, you'd want to identify clusters of these ZIP Codes. If you know that most people will drive up to three miles to buy these products, you would use three miles as the distance when calculating Gi*.

With a larger distance, you may have a few, large clusters; with a smaller distance you may have more and smaller clusters.

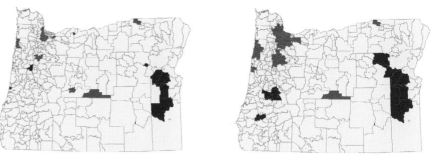

Clusters of ZIP Codes having high numbers of people more likely (orange) or less likely (blue) to buy pet supplies. Using a distance of five miles (left map), the clusters are smaller and more localized. Using a distance of 20 miles (right) creates larger, regional clusters.

Distance can be defined not only in terms of straight-line—or Euclidean—distance, but also in other ways, such as travel time, either over a street network or overland (see "Defining spatial neighborhoods and weights").

Calculating Gi*

To calculate Gi*, the GIS sums the values of the neighbors and divides by the sum of the values of all the features in the study area.

Since you're using a binary weight, based on either adjacent neighbors or neighbors within a specific distance, the attribute values are multiplied by 1 (for neighbors) and 0 (for others) so only the values of the neighbors are included. That's the same as summing the values of the neighbors.

The value of each neighbor (x_j) is multiplied by the weight for the target-neighbor pair (w_{ij}), and the results summed....

Gi* for a feature (i), at a distance (d)

$$G_i^*(d) = \frac{\sum_j w_{ij}(d)x_j}{\sum_j x_j}$$

....then the sum is divided by the sum of the values of all neighbors (x), that is, all features in the data set

Interpreting the Gi* statistic

A group of features with high Gi* values indicates a cluster or concentration of features with high attribute values.

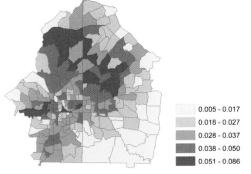

TRACT	% AGE 65+	Gi*
010103	19.08	0.033
021205	5.59	0.022
021211	7.31	0.013
021210	8.03	0.017
010204	8.18	0.013
010205	16.43	0.024
021207	13.94	0.030
021304	4.29	0.022
021212	9.43	0.023
021202	9.1	0.047
021301	6.16	0.023

0.005 - 0.017
0.018 - 0.027
0.028 - 0.037
0.038 - 0.050
0.051 - 0.086

Gi* values by census tract, for percent population age 65 and over

Conversely, a group of features with low Gi* values indicates a cold spot.

A Gi* value near 0 indicates there is no concentration of either high or low values surrounding the target feature. This occurs when the surrounding values are near the mean, or when the target feature is surrounded by a mix of high and low values.

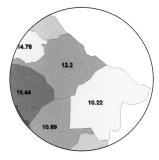

The tract having a percentage value of 10.22% (near the mean of 11.09%) and surrounded by tracts with similar values gets a low Gi* value (indicated by pale orange).

Testing the statistical significance of Gi*

As with Local Moran's I, the Z-score test is used with Gi*.

The GIS calculates the Z-score by subtracting the expected Gi* for the feature, given a random distribution, from the calculated Gi* value. The difference is then divided by the square root of the variance for all features in the study area.

The Z-score for Gi*

The expected Gi* value is subtracted from the observed Gi*....

$$Z(G_i^*) = \frac{G_i^* - E(G_i^*)}{\sqrt{Var(G_i^*)}}$$

....and the difference divided by the square root of the variance

The expected Gi* for a random distribution is equal to the sum of the weights for a given distance, divided by the number of features in the study area, minus 1.

The weights (w_j) at a distance (d) are summed....

The expected Gi* value

$$E(G_i^*) = \frac{\sum_j w_{ij}(d)}{n-1}$$

....and divided by the number of features (n), minus one

The GIS calculates a Z-score for each feature at the specified distance. As with the Gi* value itself, a high Z-score for a feature indicates its neighbors have high attribute values, and vice versa. A Z-score near 0 indicates no apparent concentration of similar values.

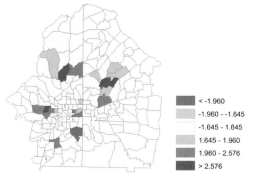

	< -1.960
	-1.960 - -1.645
	-1.645 - 1.645
	1.645 - 1.960
	1.960 - 2.576
	> 2.576

TRACT	% AGE 65+	Z-SCORE
010103	19.08	1.333
021205	5.59	-0.918
021211	7.31	-0.631
021210	8.03	-0.511
010204	8.18	-0.486
010205	16.43	0.890
021207	13.94	0.475
021304	4.29	-1.135
021212	9.43	-0.277
021202	9.1	-0.332
021301	6.16	-0.823

Census tracts color coded by Z-scores for Gi* (percent of population age 65 and over)

To determine if the Z-score is statistically significant, you compare it to the range of values for a given confidence level. For example, at a confidence level of 95%, a Z-score would have to be less than −1.96 or greater than 1.96 to be statistically significant. See "Testing statistical significance" for more on Z-scores.

Displaying the results of the Gi* statistic

As geographers Jay Lee and David Wong point out in their book *Statistical Analysis with ArcView GIS,* you can map both the individual Gi* values and the Z-scores, with the Z-scores showing you the statistical significance of the G-statistic. So, a cluster of features with low G-statistics and low Z-scores would indicate a definite cluster of low values.

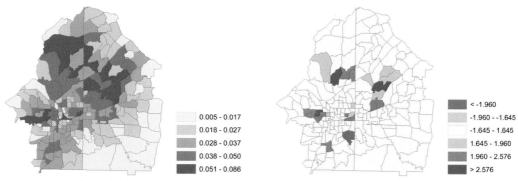

Census tracts color coded by Gi* values (left) and Z-scores, calculated from percent age 65 and over

When analyzing point features, you can symbolize the points with graduated symbols or graduated colors. Or you can generate a surface and symbolize it using a color ramp.

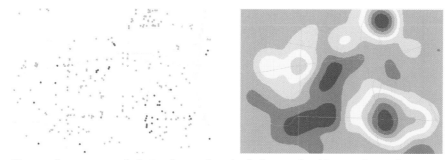

Clusters of emergency calls (using the number of calls from each address as the attribute value). Locations surrounded by locations with similarly high numbers of calls (hot spots) are shown in orange. A surface can be created from the Z-scores to present a generalized view of hot and cold spots.

For areas and raster surfaces, you can symbolize the features using graduated colors. As with Local Moran's I, you can create a map using multiple attribute values to see clusters at several distance ranges or at several confidence levels.

A more recent version of Gi*, developed by Keith Ord and Art Getis in 1995, combines the original Gi* statistic and the Z-score in a single measure. While still called Gi*, the statistic it reports is in fact a Z-score. The interpretation remains the same—high Gi* values reflect statistically significant clusters of high values, and low ones statistically significant clusters of low values.

Factors influencing the results of Gi*

Most of the same issues inherent in using Local Moran's I also apply to using the Gi* statistic.

Since the neighborhood is based on either adjacency or a specific distance, features near the edge of the study area generally will have fewer neighbors. That can skew the results for these features, since the values of the few neighbors will take on more importance in the calculation.

If you're analyzing a small number of features—fewer than 30—your results might be suspect, because of the effect of any outliers (see "Understanding data distributions").

Keith Ord and Art Getis have shown in their research that the values for Gi* are moderated when there is a strong global pattern, thus making the clusters less obvious. The local statistic works best for identifying clusters when there is no measurable pattern of clustering or dispersion across the study area. (Identifying patterns is discussed in chapter 3.)

As discussed for Local Moran's I, the Z-score is suspect when used with local statistics, including Gi*. A Bonferroni-type test may provide more valid results.

Anselin, Luc. "Local Indicators of Spatial Association—LISA." *Geographical Analysis* 27, no. 2 (1995): 93–115. Anselin lays out the conceptual basis for local spatial statistics and discusses both Local Moran's I and Local Geary's c. He also compares Local Moran's I to Gi* in the context of studying border conflict in Africa.

Getis, Arthur, and J. Keith Ord. "Local Spatial Statistics: An Overview." In *Spatial Analysis: Modelling in a GIS Environment,* edited by P. Longley and M. Batty. Wiley, 1996. Getis and Ord discuss both the Gi and Gi* statistics, as well as Local Moran's I and local statistics based on Geary's c ratio. They present a number of issues to be aware of when applying these statistics.

Lee, Jay, and David W. S. Wong. *Statistical Analysis with ArcView GIS.* Wiley, 2001. Lee and Wong provide an overview of Local Moran's I, Local Geary's c, and the Gi statistics, with illustrations of how the statistics are calculated.

Levine, Ned. *CrimeStat: A Spatial Statistics Program for the Analysis of Crime Incident Locations (v 2.0).* Ned Levine & Associates and the National Institute of Justice, 2002. The CrimeStat documentation provides an overview of clustering techniques and covers nearest neighbor hierarchical clustering in depth. Levine also discusses Local Moran's I. Includes examples of the application of the statistics to crime analysis.

Ord, J. K., and Arthur Getis. "Local Spatial Autocorrelation Statistics: Distributional Issues and an Application." *Geographical Analysis* 27, no. 4 (1995): 287–306. This paper lays out the theory and math behind the Gi and Gi* statistics. Ord and Getis also present an application of the Gi statistic to a study of the dispersion of AIDS.

Using statistics with geographic data

The statistical analysis you perform on your data tells you if there's a pattern or relationship and whether it's statistically significant. In traditional statistics, data exists in an abstract world where patterns and relationships are assumed to be devoid of external influences. But in a GIS, the features you're analyzing are associated with locations on the earth's surface. The nature of geographic data, the way the data was collected and is stored, and the choices you make about the boundaries of your study area will all influence the results of your analysis.

The nature of geographic data contradicts the basic assumptions of statistical analysis that each observation is equally likely to occur in a sample, and that observations don't influence each other. For geographic data, that would mean any feature or value would be equally likely to occur at any location, and that the presence of a feature or value wouldn't influence the occurrence of other features or values. But that's often not the case.

HOW THE NATURE OF GEOGRAPHIC DATA AFFECTS YOUR ANALYSIS

Most geographic data exhibits both regional and local trends, so features or values are not equally likely to occur at any location. If you are examining rainfall data, you might notice higher values in the eastern part of your study area with decreasing rainfall as you move west. This is a regional trend. Lower rainfall in small pockets of your study area (perhaps in valleys) is a local trend.

Rainfall in this region increases from west to east.

Similarly, in many urban areas, housing values increase as you move from flatlands to higher elevations (with views); at the same time, home prices between neighborhoods vary, and the prices of homes within a neighborhood influence each other.

Because of regional trends, how you define your study area is important. If your study area is an entire region, your analysis might show clustering of similar values, but if your study area is one neighborhood within that region, your analysis could show values as evenly dispersed.

Because of local trends, when the statistic counts features or sums values, the influence of surrounding features is included. That leads to redundancy, which can exaggerate any pattern or relationship. For example, clustering may appear to be stronger than it is.

Most spatial statistics depend on distance or some other measure of the relationship between features, such as whether they're adjacent or not. Also, in many cases, the statistics include in the calculation the number of features and the extent of the study.

Many statistics incorporate the distance between features, measured as the straight-line distance between two x,y coordinate pairs. While that's pretty straightforward if you're analyzing points (which are stored as x,y coordinate pairs), with other features the distance calculation is not as clear. Lines, for example, have to be generalized, and are often represented as a single x,y coordinate—the midpoint, an end point, a randomly selected point on the line, or the location where the lines are nearest. The software you're using may allow you to choose, or it may just pick one of the above. The distances could be very different, especially if the line is long or convoluted.

Similarly, how the lines were digitized—whether as a single continuous line or a series of short lines—will determine not only how distance is measured between lines, but also how many features are in the dataset. You may need to merge lines, or split them. While it's possible to use spatial statistics with line features, because of the difficulty in representing them as a single location, the results of the analysis may be misleading.

The results of analyzing the clustering of logging roads would vary greatly depending on the points chosen to represent the roads and whether long roads are a single line or a series of short lines.

For areas, either the centroid-to-centroid distance or the distance between nearest locations on the boundaries is used. Depending on the shape or size of the areas, these two measures could be vastly different.

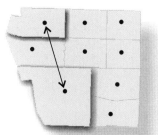 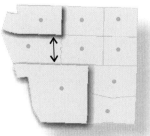

The distance can be very different depending on whether you measure from centroid-to-centroid (left) or border-to-border (right).

The calculated centroid for convoluted areas could actually be outside the area. If so, the GIS may move the coordinate to a location inside the area. In general, the statistical analysis works best if areas are of roughly equal size and shape.

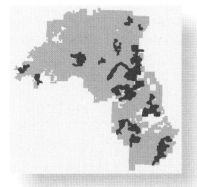

Areas with the same value for a particular attribute, such as timber harvest areas that have already been logged, may have been merged in the GIS database into single, larger areas. That will affect distance calculations as well as the number of

Some of these timber areas within a forest have likely been merged.

features in the dataset. If you're interested in the pattern formed by the timber harvest areas, you'll want to make sure you use the original units and not the merged ones.

For raster data, the main issue is the size of the grid cell. The cell size affects not only how many features are in the dataset, but also the scale at which patterns and relationships are identified. The value assigned to a cell represents the average for the cell (or for categorical data, the value that covers most of the cell). If cells are large, the data may be generalized so much that cells are very different from each other and no patterns will emerge. With a small cell size, there will be many features (cells) with the same or very similar values, and the effect of spatial dependence will be overemphasized (there will be many adjacent cells having the same value).

 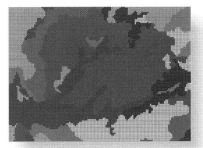

Vegetation types represented using a large cell size, and a small one

There is often more than one way to represent the same geographic feature. For example, cities could be represented as points or areas, depending on the scale of your study area. If you're working at the regional or county level, you might represent cities as areas; at the state or national level you'd represent them as points. Similarly, rivers can be represented as lines or areas.

Cities represented as areas (left), and as points (right)

Spatially continuous data (such as vegetation or rainfall) can be represented using a raster surface or using contiguous areas. The analysis results will be quite different, since both the distance calculations and the number of features will be very different. When you use a raster, you have a set of many equal-size areas, whereas with contiguous areas you'll likely have fewer areas, and they may be elongated or convoluted.

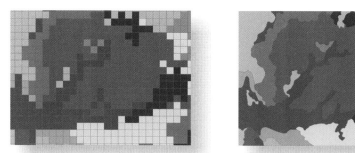

Vegetation types represented as a raster (left), and as contiguous areas (right)

The type of attribute values you're analyzing will also affect the results of your analysis. Many of the statistics are only applicable to one particular data type—nominal (categories), ordinal (ranks), or ratios, for example. So, you'll want to use the appropriate statistic, or convert the data, if possible. For example, when analyzing clusters of areas with similar values, you'll want to use ratio data, especially if the areas are of varying size, to help even out the difference in size. So, you'd use percentage of seniors in each tract instead of the number of seniors.

Because most spatial statistics include in the calculation the number of features, and some (such as the nearest neighbor index) the extent of the study area, the resulting statistical value could vary greatly depending on the study area size and configuration.

In many cases, your study area boundary will already be defined (a county or state boundary, for example). But for some studies, you may have the choice of where to draw the study area boundary. With GIS, you can define just about any study area you want. You can aggregate several areas to create a study area, create a raster covering any area you want, or select any subset of features and make the extent of the features the study area boundary.

For hierarchical areas, such as census units (blocks, block groups, census tracts, counties) or watersheds (subbasins nested into major watersheds), the attribute value you're studying will change as you move from smaller units to larger. Patterns or relationships that appear at one scale may not appear at another, because of how the data is summarized at each level.

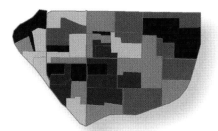 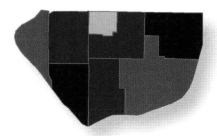

Block groups color coded by percentage of seniors (left) and census tracts color coded by the same attribute (right). Some pockets of high senior population apparent at the block group level are averaged out at the census tract level.

Much data is collected by administrative units. Because the data exists for these areas, they often become the default study area for analysis, even though they may not be the best configuration for capturing the distribution of the data you're analyzing. Events such as fires and phenomena such as air pollution don't take account of boundaries drawn by humans.

There may be features outside the boundary influencing what's going on inside the study area. You may get a more realistic picture of the processes at work in your study area if you include the surrounding area in the analysis, unless there is some barrier—such as an unbridged river— that would minimize any influence from surrounding features.

For example, if you're analyzing clustering of median house values in a county by census tract, you could buffer the county boundary to some distance, select the adjacent tracts that fall within the buffer, and calculate a statistic such as Local Moran's I for this expanded dataset. You could then show the results by mapping just the tracts within the county or, more accurately, mapping all the tracts, highlighting those inside the county.

Median house value for a county, by census tract (left) and significant clusters of similar values (right)

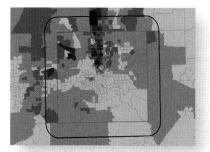

When adjacent census tracts (within the buffer) are included, the analysis identifies additional tracts that are significantly similar to their neighbors.

No matter how you define the study area boundary, features near the edge will have fewer neighbors than features in the center of the study area. This "edge effect" can make the results for features near the edge less reliable, and is particularly an issue for statistics that search for neighbors within a specified distance, such as the *K*-function.

Some solutions to this problem have been proposed, as discussed by Ned Levine in his CrimeStat documentation:

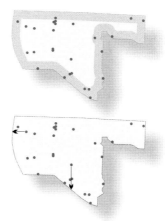

- Create a buffer within the border—the distance from features within the buffer to other features is not calculated, but the distance is calculated to the features within the buffer from features in the interior of the study area.

- Assign weights to features closer to the boundary; the closer to the boundary, the higher the weight. The weight is then multiplied by the number of neighboring features within the distance of each feature, to account for any assumed neighboring features outside the boundary.

These solutions work best with social and cultural (human-made) boundaries where there are likely to be features outside the boundary.

GIS DATA **AND ERROR**

Error is a problem for all data. But because GIS data attempts to represent objects that have a geographic location and characteristics such as width or area, there is additional potential for error to creep into the analysis. There may be errors in the way the data was collected or digitized, or even in the way it is represented.

The location may be wrong, data features that exist may have been missed, or ones that don't exist incorrectly included in the dataset. The data may have been generalized, so that small differences or subtleties that would have been important for analysis will have been lost. Similarly, the data may have been classified incorrectly, perhaps receiving the incorrect category, or an incorrect numeric value.

And errors can be propagated in the analysis itself. This is especially true in spatial statistics, where distances between features are calculated, and values of features are compared to each other or to mean values for the dataset. In many cases the results of the comparison are then multiplied by a weight value before being summed. A few minor errors can be magnified in the process.

To the extent you're confident in the quality of your GIS data, you can be confident in the quality of your analysis results. The results of your analysis are just another piece of information you can use to make and support decisions. They should be considered along with your knowledge of the study area and of the geographic features or phenomenon you're analyzing. Undoubtedly, political and economic factors will also enter into the decision-making process.

References

Anselin, Luc, and Arthur Getis. "Spatial Statistical Analysis and Geographic Information Systems." *The Annals of Regional Science* 26 (1992): 19–33.

Anselin, Luc, Rustin F. Dodson, and Sheri Hudak. "Linking GIS and Spatial Data Analysis in Practice." *Geographical Systems* 1 (1993): 3–23.

Bailey, Trevor C., and Anthony C. Gatrell. *Interactive Spatial Data Analysis.* Longman, 1995.

Goodchild, Michael, et al. "Integrating GIS and Spatial Data Analysis: Problems and Possibilities." *International Journal of Geographical Information Systems* 6, no. 5 (1992): 407–23.

Getis, Arthur. "Spatial Statistics." In *Geographical Information Systems: Principals, Techniques, Management and Applications,* edited by P. Longley, M. F. Goodchild, D. J. Maguire, and R. W. Rhind, 239–51. Wiley, 1999.

5 Analyzing geographic relationships

Analyzing the relationships between geographic phenomena gives you a better understanding of what's happening in a location, lets you predict where something might occur, and helps you examine why things occur where they do.

In this chapter:

- Why analyze geographic relationships?
- Using statistics to analyze relationships
- Identifying geographic relationships
- Analyzing geographic processes

Beyond analyzing how geographic features are distributed, GIS analysis can be used to analyze the relationships between features. Identifying and measuring relationships lets you better understand what's going on in a place, predict where something is likely to occur, or begin to examine why things occur where they do.

UNDERSTANDING WHAT'S GOING ON IN A PLACE

A transportation analyst working to reduce traffic accidents across a county could analyze the relationship between accidents and speed limit on highways. Showing that accidents increase as the speed limit increases could help officials make the case for reducing the speed limit on certain sections of highway.

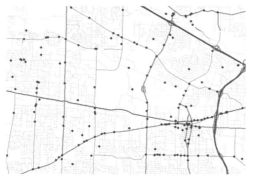

Traffic accidents, with streets color coded by speed limit (the darker the color, the higher the limit)

Of course, there are other factors involved—traffic accidents are related to traffic volume, weather conditions, type of paving, and so on, in addition to speed limit. One of the benefits of analyzing relationships is the ability to hold other factors constant and explore the relationship between specific factors, all other things being equal.

An environmental lawyer representing low-income residents of a city could analyze toxic sites in poor neighborhoods. If high-income neighborhoods tend to have fewer and low-income neighborhoods more—all other things being equal—the lawyer might make a case that the residents of the low-income neighborhoods are facing a form of discrimination.

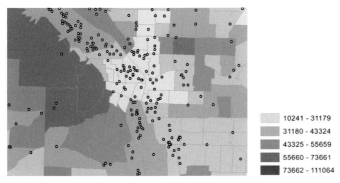

	10241 - 31179
	31180 - 43324
	43325 - 55659
	55660 - 73661
	73662 - 111064

Locations of toxic sites, with census tracts color coded by median income

PREDICTING WHERE SOMETHING IS LIKELY TO OCCUR

An archaeologist with limited funds for research would want to identify the locations where artifacts are most likely to be found within a study area. She'd analyze the landforms and soils that known sites occur in. She'd look for that combination of landforms or soils in other parts of the study area, and focus her excavations in those areas.

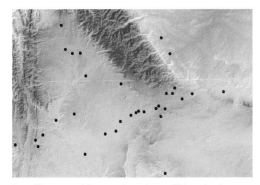

Landforms and known locations of historic sites

An extractive industries company wanting to reduce exploration costs and minimize damage to the environment would try to predict where certain minerals will be found by comparing known occurrences of the minerals to the geology in those spots, nearby locations of oil and natural gas beds, and other factors. By determining which factors are most related to the minerals, they can look for other occurrences of these factors and explore for the minerals in those locations.

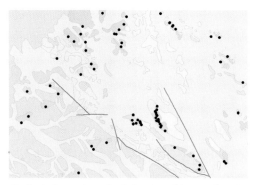

Geologic formations, fault lines, and known locations of minerals.

EXAMINING WHY THINGS OCCUR WHERE THEY DO

A state public health agency wishing to improve the health of newborns could analyze the percentage of low birth weights by county in conjunction with other health, environmental, economic, and demographic variables. It might be that certain factors—such as the educational or income level in a county—are related to a high percentage of babies with low birth weight. The information would be useful in addressing the underlying causes through educational outreach or preventive health care in those counties.

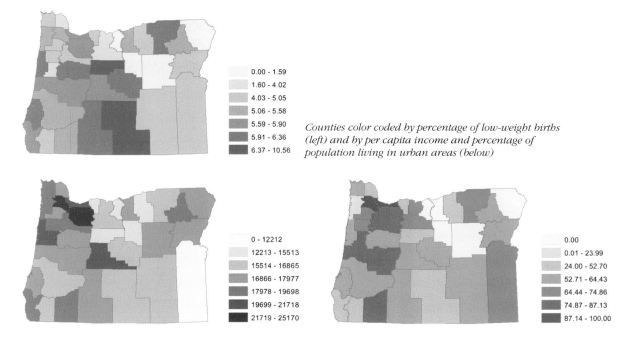

0.00 - 1.59
1.60 - 4.02
4.03 - 5.05
5.06 - 5.58
5.59 - 5.90
5.91 - 6.36
6.37 - 10.56

Counties color coded by percentage of low-weight births (left) and by per capita income and percentage of population living in urban areas (below)

0 - 12212
12213 - 15513
15514 - 16865
16866 - 17977
17978 - 19698
19699 - 21718
21719 - 25170

0.00
0.01 - 23.99
24.00 - 52.70
52.71 - 64.43
64.44 - 74.86
74.87 - 87.13
87.14 - 100.00

A wildlife biologist trying to determine what constitutes preferred habitat for a particular species could analyze the conditions—such as climate, elevation, vegetation, food supply, proximity to water, and so on—where the species occurs. Certain factors may be more important than others for the species to survive. The information would be useful in protecting critical habitat for the species.

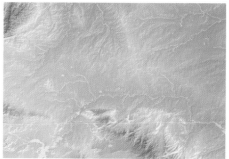

Bobcat density (top), elevation, and land cover

When you look for relationships, chances are you've already formed an opinion, or at least a question, about whether a relationship exists. Your suspicions may be based on your knowledge of the phenomena you're studying or on a visual analysis of maps of the attributes you suspect may be related. Using statistics allows you to verify a relationship and measure how strong it is.

THE IDEA BEHIND USING STATISTICS TO ANALYZE RELATIONSHIPS

The basic idea behind using statistics to measure relationships is seeing to what extent the value of one attribute changes when the value of another attribute changes. In statistics, attributes are referred to as variables. You can think of the analysis as measuring the relationship between two or more map layers representing the variables. If you're interested in the relationship between where senior citizens live and home value, conceptually you'd analyze the relationship between a layer of census tracts showing percentage of population age 65 and over and a layer of the census tracts representing median home value.

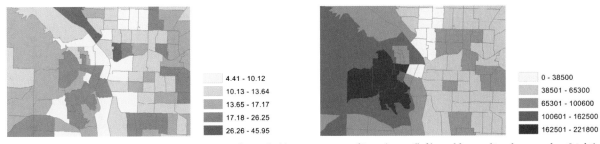

4.41 - 10.12		0 - 38500
10.13 - 13.64		38501 - 65300
13.65 - 17.17		65301 - 100600
17.18 - 26.25		100601 - 162500
26.26 - 45.95		162501 - 221800

Census tracts color coded by percent age 65 and over (left) and by median home value (right) represent two layers of data.

In fact, you're analyzing the relationship of two or more attribute values associated with each feature—a sample point, a defined area, or a raster cell—representing a unique piece of geography and often stored in a single table. Using the above example, you'd analyze the relationship between the variables "percentage of people age 65 and over" and "median home value" in each census tract.

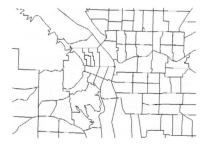

TRACT	PERCENT 65+	MEDIAN VALUE
0043	12.68	66600
0070	15.02	143500
003601	10.73	36900
003902	19.95	49400
003602	12.02	47800
0044	45.95	68800
003603	24.19	54000

Percent age 65 and over and median home value are variables associated with census tracts.

The statistical methods presented in previous chapters looked at the distribution of a single attribute associated with the features in a single layer. Statisticians call this type of analysis "univariate." When measuring relationships, you work with two attributes (bivariate analysis), or several (multivariate analysis).

ASSIGNING VARIABLES TO GEOGRAPHY

To analyze relationships using GIS, you need to make sure the variables from the different layers are associated with the same geographic unit. If you're analyzing attributes associated with a set of administrative units, such as counties or census tracts, this will already be the case, or you can easily join attributes in a table to the existing geographic units, as long as the table includes an ID for each unit.

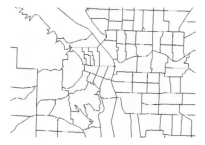

TRACT	PERCENT 65+	MEDIAN VALUE
0043	12.68	66600
0070	15.02	143500
003601	10.73	36900
003902	19.95	49400
003602	12.02	47800
0044	45.95	68800
003603	24.19	54000

TRACT	MEDIAN INCOME
0043	51705
0070	108931
003601	33869
003902	46767
003602	43788
0044	58750
003603	51447

Additional variables (median income, in this example) can be joined to an existing table.

If the areas are of different sizes, you'll want to account for this by using a ratio. There may be more emergency calls in a particular block group simply because that block group is bigger or has more people living in it. So, you'd use a density value (calls per square mile) or a proportion (calls per 100 people).

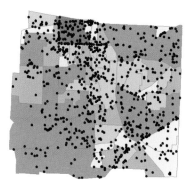

BLOCK GROUP	POPULATION	CALLS	CALLS PER 100
410670317021	1046	33	3.154880
410670312002	5774	145	2.511260
410670312001	246	18	7.317070
410670311002	619	19	3.069470
410670304021	3714	57	1.534730
410670304012	3434	66	1.921960
410670311001	1743	52	2.983360

Locations of emergency calls and block groups color coded by population

For attributes associated with different sets of features, you'll need to combine the features—one way or another—so that the attributes refer to the same geographic location or unit. For example, if you're analyzing the point locations of moose sightings and land cover delineated as area features, you'd assign the land-cover type to each point using a point-in-polygon overlay routine.

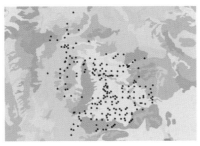

OBSERVATION ID	QUANTITY	LAND COVER
000000017100	1	42001
000000018100	1	42004
000000604201	1	42004
000003789101	1	42001
000010015501	36	32006
000010018201	15	32006
000010113101	1	42004

Overlaying locations of moose sightings and land-cover areas adds a land-cover code to each location.

If you need to combine two or more sets of area features, you can either do a polygon overlay or create rasters of the areas using the same origin and cell size for each raster. You can then join the attributes to a single raster. If you have many areas or they are convoluted, it may be easier to work with the data by converting it to raster—polygon overlay will create slivers and areas of varying size.

A special case exists when you want to analyze sets of points representing different categories of features. For example, a crime analyst might want to see if there's a relationship between the locations of liquor stores and assaults. In this case, the analyst would need to summarize the number of features by area (the number of liquor stores and assaults in each block group or raster cell) and use these counts as the variables to analyze.

You can use a similar method to compare the location of events with other attributes. For example, if you wanted to know if there is a relationship between emergency calls and the economic status of neighborhoods, you'd count the number of emergency calls in each neighborhood and see if there is a relationship to some measure of economic status, such as median income or median house value associated with each neighborhood. Of course, the number of calls is partly dependent on the number of people in the neighborhood, so you'd want to account for this by using calls per person rather than the raw count.

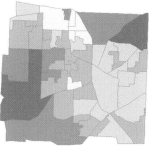

BLOCK GROUP	MEDIAN VALUE	CALLS PER 100
410670317021	76400	3.154880
410670312002	63900	2.511260
410670312001	100000	7.317070
410670311002	54300	3.069470
410670304021	141200	1.534730
410670304012	99800	1.921960
410670311001	58900	2.983360

Locations of medical emergency calls and block group boundaries (left) and block groups color coded by median house value (right)

You can also create variables that represent the spatial interaction between features, such as distance, travel time, or travel cost. A biologist evaluating wildlife habitat could create a surface of distance from streams and include the values as one of several variables associated with each observation.

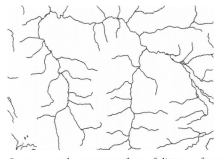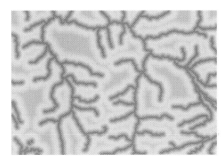

High : 2816.0 m

Low : 0.0 m

Streams, and a raster surface of distance from streams

USING STATISTICS TO ANALYZE GEOGRAPHIC RELATIONSHIPS

Statistical methods for measuring relationships are well established in traditional (nonspatial) statistics. There are a number of assumptions about the data these methods require for the results to be valid. Two key ones are that any value is equally likely to occur in a sample, and that the value of one observation does not affect the value of any other. Geographic data, however, often doesn't hold to these assumptions.

Attribute values vary across a region

You may find a regional or directional trend for one or more of your variables. For example, if one of your variables is rainfall, you may find a trend of higher to lower rainfall as you move from east to west.

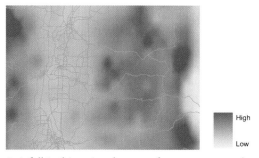

High

Low

Rainfall in this region decreases from east to west, then increases again.

The assumption of statistics is that the relationship is constant across the study area. But regional trends—usually caused by some underlying factor or factors—influence attribute values, so you need to account for them before you can determine if there are relationships between variables. Often, an extra variable is included—such as one to indicate whether each rainfall value occurs in the eastern or western part of the study area.

You can check for regional variation using a trend plot, available in GIS software packages such as ArcGIS.

Nearby features are more similar than distant ones

Geographic features that are near each other are likely to be more similar than distant features, a phenomenon referred to as spatial autocorrelation. Spatial autocorrelation violates the assumption that observations are independent. It doesn't matter whether the influence is direct (the assessed value of a house being partly dependent on the value of surrounding houses) or indirect (adjacent fields having high crop yields because they all lie on nutrient-rich soils). In essence, spatial autocorrelation introduces redundancy into the analysis, since the influence of nearby features with similar values is reflected in the value of each feature. That leads to overcounting when statistical methods are used—if your sample includes many such features, those values will be amplified in the analysis, and your results won't be accurate.

Spatial autocorrelation is partly a function of the geographic units you're using. Small units (such as census block groups) are more likely to be similar to their neighbors than are large units (such as counties). Larger units may mask local variation since values are summarized across the entire unit. Population may be concentrated in one corner of a county, but the density value will be calculated as though the population was spread evenly across the county.

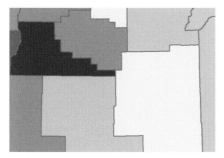 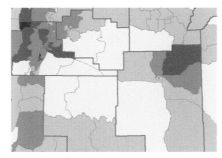

Median house value by county and by census block group. While neighboring counties have different values, many neighboring block groups have similar values.

You can check for spatial autocorrelation in your data using one of the methods discussed in chapter 3, such as Moran's I or Geary's c.

Researchers have proposed several techniques for dealing with the issues inherent in geographic data when measuring relationships. Broadly, the techniques involve either resampling your data until you get a sample that no longer violates the assumptions, isolating the spatial components of the relationships so the analysis can be performed without the spatial influences that violate the assumptions, or incorporating the spatial influences into the analysis so the results more accurately reflect the real-world relationships.

IDENTIFYING RELATIONSHIPS VERSUS ANALYZING PROCESSES

You may simply want to know whether there is a relationship between two variables and the nature of the relationship. To do this, you measure the extent to which two variables vary together—that is, as one changes in value, the other changes proportionally (known as covariation). By identifying the relationship, you can take some action or simply get a better understanding of what's happening in a place.

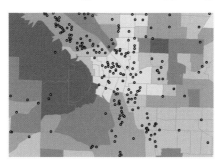

Locations of toxic sites, and census tracts color coded by median income

Or, you may want to analyze a complex process to find out which variables are driving the process or to predict values of a variable. If so, you'd analyze the extent to which a change in value of a particular variable depends on a change in value of other variables.

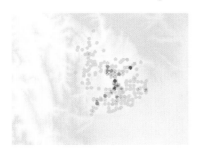  

Moose density, slope, and distance to streams

Dependence implies a stronger relationship between variables. With covariation, the relationship is reciprocal—as one value changes, the other changes proportionally, and vice versa. With dependence, the relationship is not necessarily reciprocal—if the value of one variable depends on another, the reverse may not be true.

Identifying geographic relationships is based on the idea that you can measure how much two attributes for a given piece of geography vary in the same way. To the extent they do, there is an apparent relationship between them.

You can see how two variables relate to each other by displaying them on the kind of chart known as a scatter plot.

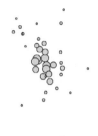

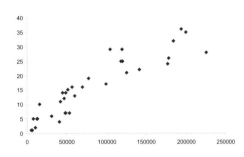

Grocery stores symbolized by population within three miles—stores in densely populated areas have more nearby competitors. The scatter plot shows the relationship between these variables.

STORE #	COMPETITORS	POPULATION
205	1	6504
214	13	59393
43	16	56145
221	2	10314
18	5	12813
219	10	16019
207	16	69142

If the values of one variable increase as the other increases, there is a direct relationship. If the values of one decrease as the other increases, the relationship is inverse. Otherwise, there is no relationship. The statistical term for this type of relationship is correlation. If two variables have a direct relationship there is said to be positive correlation. Two variables with an inverse relationship display negative correlation.

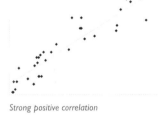

Strong positive correlation

Weak negative correlation

Positive and negative describe the direction of the correlation. If you drew a straight line on the scatter plot between all the data points, you could see the direction of the correlation. The extent to which the points are clustered or dispersed around the line indicate the strength of the correlation. If all the points lay on the line (not likely for most datasets), the correlation would be perfect.

While you might suspect a relationship—and be able to see it visually on a scatter plot—using statistics to measure the relationship lets you confirm that there is a relationship and measure its direction and strength. Once you've calculated a statistic, you can test it to see if the relationship is significant so you can be more confident in any decisions you make based on the analysis.

METHODS OF IDENTIFYING GEOGRAPHIC RELATIONSHIPS

A commonly used method for variables that are interval or ratio values is Pearson's correlation coefficient, developed by British mathematician Karl Pearson in the early 1900s.

A different method, Spearman's rank correlation coefficient (developed by British psychologist Charles Spearman, also in the early 1900s), is used with ordinal (ranked) values. You'd use Spearman's rank correlation coefficient if you were analyzing the preferences of different groups of people for restaurants in an entertainment district. You'd ask people to rank the restaurants in order of preference, then see if, for example, the preferences of seniors were negatively correlated with the preferences of people who have small children.

USING PEARSON'S CORRELATION COEFFICIENT

Pearson's correlation coefficient is the ratio of the joint variation of two variables to the total variation of the entire dataset. The numerator is the covariance of the two variables—the extent to which an increase in one results in a proportional increase (or decrease, for negative correlation) in the other. If the covariance is 0 (indicating no relationship), the ratio will be 0.

The denominator represents the total variation in the dataset, calculated by multiplying the standard deviations of the two variables.

The values of the two variables (x and y) for each feature are multiplied, and the products summed and divided by the number of features (n)....

Pearson's correlation coefficient

$$r = \frac{\dfrac{\sum_{i} x_i y_i}{n} - \bar{x}\,\bar{y}}{s_x s_y}$$

....then the product of the mean value of the variables is subtracted from the result.

The numerator (representing the covariance) is divided by the variation in the data set as a whole, represented by the product of the standard deviations of the two variables.

The value of the correlation coefficient (r) ranges from 1 (indicating a perfect direct relationship) to −1 (a perfect inverse relationship). A value of r that approaches these values indicates that almost all of the variation in the dataset is accounted for by the covariance of the two variables you're analyzing. A value near 0 for r indicates the covariance in the two variables is negligible compared to the variation for the total dataset.

	0.20 - 0.82
	0.83 - 1.36
	1.37 - 2.36
	2.37 - 4.05
	4.06 - 7.32

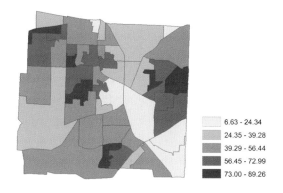

	6.63 - 24.34
	24.35 - 39.28
	39.29 - 56.44
	56.45 - 72.99
	73.00 - 89.26

Number of medical emergency calls per 100 people (left) and percentage of single-family housing in each block group (right). The correlation coefficient for these two variables in this area is −.65, a moderate inverse relationship. As single-family housing decreases, the number of calls increases, as seen on the scatter plot.

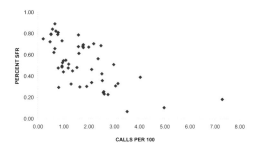

The null hypothesis for the significance test is that the correlation coefficient is 0—that there is no correlation between the variables. To test the significance of the coefficient, you use a t-test. You calculate the value for t and compare it to the critical value obtained from a standard table for Student's t.

r is multiplied by the square root of....

The t score

$$t = r\sqrt{\frac{n-2}{1-r^2}}$$ *....the number of features (n) minus two, divided by one minus r-squared.*

In most cases, you'll be interested in whether the correlation is significant in a particular direction (either direct or inverse), so you'll use a one-tailed test. When performing a one-tailed test, the null hypothesis includes the direction—for example, that there is no inverse correlation between the variables. If you don't care whether the relationship is direct or inverse but only whether there is a relationship, you'd use a two-tailed test. The critical value will differ depending on whether you're using a one-tailed or two-tailed test. Specifically, the confidence level for a one-tailed test is half that for a two-tailed test. For a sample of at least 120 features, the critical value for a two-tailed test at a confidence level of 0.05 is 1.96—equal to the critical value for a one-tailed test at a confidence level of 0.025.

You also need to know the degrees of freedom in order to obtain the correct critical value (see the discussion of the Chi-square test in chapter 3 for more on degrees of freedom). When using Pearson's correlation coefficient, the degrees of freedom is always the number of features minus 2. One degree of freedom is lost because only $n-1$ features is needed to explain all the variation. Once you know the variation for all other features, the last one is redundant—you can figure out its variation from the already calculated variation. Another degree of freedom is lost by taking the variation around the means for the two variables.

In the example above, there are 52 block groups, so the degrees of freedom is 50. The calculated t-value of 6.5 exceeds the critical value for t of 2.39 at the 0.01 confidence level (for a one-tailed test). There is less than a 1% chance you'd be wrong in not accepting the null hypothesis—the correlation is statistically significant at the 0.01 level.

One of the assumptions for testing Pearson's correlation coefficient is that both variables are from normally distributed populations (see "Understanding data distributions" and "Testing statistical significance"). If that's not the case—and it often isn't with geographic data—the results of the analysis may be biased.

One option is to transform your data. That will convert the values to a new scale, changing the distribution. Depending on the transformation and the parameters you use, you may be able to obtain a distribution of values that is close to normal. You can then use Pearson's correlation coefficient. Several of the references at the end of the chapter, including Lawrence Hamilton's *Regression with Graphics,* discuss data transformations.

Another option is to convert the interval data to ordinal (ranked) data, and use Spearman's rank correlation coefficient. For example, suppose a public health official wanted to see if counties having many children born with low birth weights also had many people living in rural areas. By using a QQ normal plot, she could check to see if both variables are normally distributed.

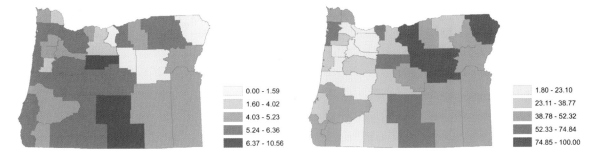

0.00 - 1.59	1.80 - 23.10
1.60 - 4.02	23.11 - 38.77
4.03 - 5.23	38.78 - 52.32
5.24 - 6.36	52.33 - 74.84
6.37 - 10.56	74.85 - 100.00

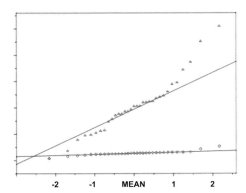

Counties color coded by percentage of births that are low weight and by percentage of the population living in rural areas. The line on the QQ plot shows expected values for a normal distribution—the closer the values to the line, the closer the distribution is to normal. The plot shows that while the rural percentage values are normally distributed (lower line), the low-birth-weight percentage values are not.

If the variables are not normally distributed, she could rank the counties by both variables and use Spearman's rank coefficient to measure the correlation.

COUNTY	% LOW WEIGHT	LOW WEIGHT RANK	% RURAL	RURAL RANK
41009	3.63	31	44.96	14
41007	4.74	26	41.45	18
41059	6.36	3	31.43	27
41063	0	34	100	1
41049	5	22	47.37	13
41061	5.50	17	41.41	19
41021	5.56	16	90.42	5
41057	5.49	18	74.84	6

Counties labeled with their IDs. The table shows each county's rank for percentage of births that are low weight, and percentage of the population living in rural areas.

The danger in doing this, though, is that by converting interval or ratio data to ranks you lose the magnitude of the variables, and therefore you'll have less confidence in the results of your analysis. For example, if you ranked counties by percentage of rural population, the top four values may be bunched but the fourth one still gets a rank of four; in the other set of values you're comparing (for example, percentage of births that are low weight), the value ranked fourth may be much lower than the value ranked first.

COUNTY	% RURAL	RURAL RANK
41063	100.00	1
41055	100.00	2
41069	100.00	3
41023	100.00	4
41021	90.42	5
41057	74.84	6
41031	68.90	7

COUNTY	% LOW WEIGHT	LOW WEIGHT RANK
41031	10.56	1
41037	8.57	2
41059	6.36	3
41041	6.24	4
41013	6.20	5
41035	6.18	6
41043	6.07	7

The top four counties all have the same value for percentage of rural population, but there is a range of values for the top four counties by percentage of low-weight births.

USING SPEARMAN'S RANK CORRELATION COEFFICIENT

Spearman's rank correlation coefficient measures the extent to which two lists of ranked values correspond. The coefficient is based on the difference in rank between each feature for the two variables. The difference is squared (to make all the values positive) and summed. The sum is multiplied by 6 and the result divided by the number of features (n) cubed, minus the number of features (or, in another form of the equation, the number of features minus 1 multiplied by the number of features squared—the value is the same). This is then subtracted from 1 to get the value of r_s.

Spearman's rank correlation coefficient

The difference (D) between rank values for each feature is squared, the squares for all features are summed, and multiplied by 6....

$$r_s = 1 - \frac{6\Sigma D^2}{n^3 - n}$$

....the result is divided by the number of features (n) cubed minus the number of features, and the result of this ratio subtracted from one.

As with Pearson's correlation coefficient, the value of r_s ranges from 1 (a perfect direct correlation) to −1 (a perfect inverse correlation). A value near 0 indicates that the two set of ranks have no relationship.

COUNTY	RURAL RANK	LOW WEIGHT RANK
41063	1	34
41055	2	35
41069	3	36
41023	4	33
41021	5	16
41057	6	18
41031	7	1
41037	8	2

In this example, r$_s$ is −0.30, indicating a weak inverse correlation—as rural population increases (left), there is a slight drop in the percentage of low-weight births.

To test the statistical significance of the results, a t-test can be used, and the value of t compared to a critical value obtained from a standard table for Student's t. As with Pearson's coefficient, the degrees of freedom is the number of features minus 2, and you can perform either a one-tailed or two-tailed test, depending on whether your null hypothesis includes the direction of the correlation (direct or inverse) or simply whether there is a correlation. In most cases, you'll perform a one-tailed test.

WHAT THE CORRELATION COEFFICIENT DOESN'T MEASURE

Any relationships that you identify apply only to the geographic location and the scale of the study area for the analysis you're performing. You couldn't, for example, apply the results of the correlation for one county to other counties in the state or country. Similarly, if a correlation exists for census tracts within a county, you couldn't apply the results to block groups within the same county.

The correlation coefficient does not measure causation—that is, whether an increase in the value of one variable causes an increase in the value of the other. It measures only covariation—the extent to which the values of the variables vary at the same rate. That doesn't mean causation doesn't exist, only that the correlation coefficient cannot tell you if it does. In *Geographic Measurement and Quantitative Analysis,* Robert Earickson and John Harlin list a number of possible reasons correlation exists outside of direct causation. These include an indirect cause affecting both variables; extreme values in one or both variables that make it appear correlation exists; lack of independence between the variables, such as when spatial autocorrelation is present; and chance.

Correlation analysis also doesn't tell you why there is a relationship (although you may suspect the underlying factors)—it simply tells you the variables are related.

While correlation measures the strength and direction of a relationship, it doesn't measure the form of the relationship. That is, it simply measures the dispersion of points around a straight line. If the relationship takes a form other than a straight line, correlation analysis will either not be able to measure the relationship, or it will give inaccurate results. The relationship might be nonlinear—as the value of one variable increases, the value of the other increases or decreases at a faster rate. A scatter plot of the data you're analyzing will give you a sense of the form of the relationship between the variables, and thus whether or not you can use correlation for your analysis.

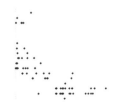

Curvilinear relationship

People analyze geographic processes to be able to predict the likelihood that something will occur in a place. For example, a retail analyst might want to predict store sales using the size of the store, the number of competitors, and the number of potential customers in the vicinity. By plugging in the values at a particular location, the analyst could predict sales for a store built there.

STORE	SQUARE FEET	COMPETITORS	POPULATION
205	28000	1	6504
214	39000	13	59393
43	30000	16	56145
221	60000	2	10314
18	35000	5	12813
219	34000	10	16019
207	65000	16	69142
208	34000	14	48125

Locations of grocery stores, and associated variables

Another reason people analyze geographic processes is to identify the underlying factors. For example, a wildlife biologist may want to identify the most important components of the habitat for a species—things like elevation, distance to water, and land cover. Knowing that could help policy makers design legislation to protect the species' habitat.

Moose sightings with elevation, distance to water, and land cover

OBSERVATION ID	ELEVATION	DISTANCE TO WATER	LAND COVER
36	2195	750	42001
42	2140	250	42004
938	1916	250	42004
3616	2031	250	42001
6805	2242	500	32006
6826	2072	1000	32006

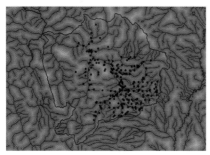

Whatever your reason, you start by developing a theory as to what's driving the process. You then analyze the relationships between various attributes of your data, representing those components—the variables. The definition of the relationships is often referred to as a model.

One method for analyzing the relationships between variables is regression analysis. Regression analysis, like other types of statistical modeling, can get quite complex. People sometimes spend years developing and refining their models, trying to capture the ways in which variables interact. Nonetheless, even a simple regression analysis can provide additional information for your decision-making process—as long as you make sure the results are valid.

LINEAR REGRESSION ANALYSIS

One type of regression analysis, known as linear regression, is a common approach for building simple models to analyze geographic processes. As with correlation analysis, a pair of values for each feature can be plotted as a data point on a chart. If you plotted all the pairs, you'd have a graphic representation of the relationship. The idea behind linear regression is to find the best fit of a line between the data points on the chart—that line represents the relationship. There are a number of ways to fit the line to the points. One common method is called ordinary least squares (OLS), which minimizes the squared distance (in data space) from the points to the line, measured parallel to the y-axis. All the distances are measured, squared, and summed. The line that minimizes the sum is the best fit.

However, not all relationships are linear. This is especially true for geographic relationships. If you're analyzing the relationship between the number of driving trips people make to a shopping mall and how far away they live, you'd likely find that the number of trips drops off sharply as distance increases (up to a couple miles), then levels off. If you created a chart with distance on the x-axis and number of trips on the y-axis, you could see graphically that the relationship forms a curved line rather than a straight one. Nonlinear regression, which is beyond the scope of this book, deals with these types of relationships.

For ordinary least squares regression, the variable you're predicting (the dependent variable) and the variables you're using to predict (the independent variables) must be interval or ratio values.

Nominal (categorical) data can't be used unless you effectively convert it to interval or ratio data. To do this, you create an attribute for each possible category, and assign binary values—a 1 indicates that the feature is in a category, and a 0 indicates it isn't in the category. These are sometimes referred to as dummy variables. If one of your variables is land-cover type and the values are forest, wetland, shrub, and so on, you'd add an attribute named "Forest" to the layer's attribute table. A value of 1 would indicate that a feature is forest, a value of 0 that it isn't. You'd do the same for wetland, shrub, and the rest of the land-cover types.

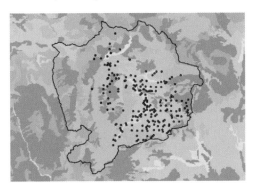

OBSERVATION ID	LAND COVER	NAME	GRASSLAND	SHRUB	WOODLAND	FOREST
36	42001	Woodland	0	0	1	0
42	42004	Forest	0	0	0	1
938	42004	Forest	0	0	0	1
3616	42001	Woodland	0	0	1	0
6805	32006	Grassland	1	0	0	0
6826	32006	Grassland	1	0	0	0
7352	42004	Forest	0	0	0	1

Moose sightings with land cover. Each land-cover type is represented as a dummy variable.

How ordinary least squares regression works

In its simplest form, ordinary least squares regression shows the relationship between two variables—the independent variable, x, which you're using to predict, and the dependent variable, y, which is the one you want to predict. This is known as bivariate regression. If you can define the relationship between y and x, using the observed values, then you could get the predicted value of y for any value of x. Suppose a retail analyst wants to analyze the factors contributing to store sales, and suspects that sales depend—at least partly—on the size of the store. The model would be set up so that the dependent variable (y) is store sales, and the independent variable is the square footage of the store. The analyst would define the relationship using the sales and size of existing stores. Once the relationship is established, the analyst could determine to what degree sales depend on store size.

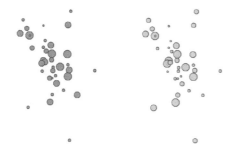

STORE	SALES	SQUARE FEET
205	13700	28000
214	32000	39000
43	13700	30000
221	17700	60000
18	27000	35000
219	22000	34000
207	27000	65000

Grocery stores symbolized by sales (left) and size in square feet.

The relationship can be expressed as an equation that defines the best fit for the line.

The predicted value of the dependent variable

The value of an observation (i) of the independent variable (x) is multiplied by the slope (b)....

$$\hat{y}_i = a + bx_i$$

....and added to the intercept (a)

In the equation, *a* and *b* are known as regression coefficients—*a* is where the line intersects the y-axis (the intercept) and *b* is the slope. By multiplying any observed value of *x* by the slope and adding to it the intercept, you get the predicted value for *y* (often shown as ŷ, pronounced "y hat," to distinguish it from an observed *y* value).

The value for *b* (the slope) is calculated from the observed *x* and *y* values.

For each observation (i), the values of the independent variable (x) and the dependent variable (y) are subtracted from their respective means, and the differences multiuplied; the products are summed for all observations....

The slope coefficient

$$b = \frac{\sum_i (x_i - \bar{x})(y_i - \bar{y})}{\sum_i (x_i - \bar{x})^2}$$

....the sum is then divided by the sum of the squared differences between each independent variable and the mean.

Then *a* can be calculated from *b*.

The intercept coefficient

The slope coefficient (b) is multiplied by the mean of the independent variable....

$$a = \bar{y} - b\bar{x}$$

....and the result subtracted from the mean of the dependent variable

Once you know the values of *a* and *b*, you've defined the line that represents the relationship of the variables. You can plug in the observed *x* values and calculate the predicted *y* for each *x*.

Interpreting the results of regression analysis

By comparing the variance in the predicted values to the variance in the observed values, you can tell how well your model is working.

The coefficient of determination

$$r^2 = \frac{s^2_{\hat{y}}}{s^2_y}$$

The variance in the predicted values is divided by the variance in the known values of the dependent variable

With a perfect fit—all the observed values the same as the predicted values and lying on the line—the variances would be equal, and r^2 would equal 1. Anything less than a perfect fit (which is the likely outcome) would result in values less than 1.0. An r^2 of 0.80, for example, means that 80% of the variance of the dependent variable is accounted for by the variation in the independent variable. The closer to 1, the more dependence there is.

STORE	SALES	SQUARE FEET
205	13700	28000
214	32000	39000
43	13700	30000
221	17700	60000
18	27000	35000
219	22000	34000
207	27000	65000

With store sales as the dependent variable and store size (in square feet) as the independent variable, the r² for these stores is 0.67—about two-thirds of the variation in store sales is explained by variation in the size of the stores.

The relationship is not necessarily reciprocal—you couldn't necessarily predict the square footage of a store from the sales.

You'll also want to calculate the differences between the observed y values and the predicted *y* values—that is, for each, what the model predicted for the value of *y*, and the observed value of *y*. These differences are termed residuals. Calculating the residuals is necessary for testing the validity of your model.

STORE	SALES	PREDICTED SALES	RESIDUALS
205	13700	17960	-4260
214	32000	21533	10467
43	13700	18610	-4910
221	17700	28355	-10655
18	27000	20234	6766
219	22000	19909	2091
207	27000	29979	-2979
208	13700	19909	-6209
218	22000	18610	3390

The known sales minus the predicted sales for each store is the residual value.

Using more than one independent variable

Most geographic processes aren't controlled by a single variable. So, the retail analyst might decide that while store size is a pretty good predictor of sales, combining store size with additional factors, such as number of nearby competitors and number of potential customers in the vicinity, will improve the prediction. The model would now include three independent variables. A regression analysis with two or more independent variables is called multivariate regression.

A multivariate regression—each independent (x) variable has an associated coefficient

$$\hat{y} = a + b_1 x_1 + b_2 x_2 + b_3 x_3$$

The r^2 for the multiple regression describes the variation in y explained by the combination of independent variables. There is still a single set of observed y values and a set of predicted y values, but in this case, the y values are predicted from several independent variables that are used to calculate r^2.

STORE	SQUARE FEET	COMPETITORS	POPULATION
205	28000	1	6504
214	39000	13	59393
43	30000	16	56145
221	60000	2	10314
18	35000	5	12813
219	34000	10	16019
207	65000	16	69142
208	34000	14	48125

Including the number of competitors for each store and the population within three miles, as well as the store size, improves the regression model slightly, resulting in an r² of 0.69.

Identifying the key variables

If the goal of your analysis is to find out which variables are driving the process, you'd test the significance of each variable's correlation coefficient *(b)* using a t-test. A planner interested in identifying the factors determining where calls for medical assistance are likely to occur would include such variables as the percentage of a particluar type of land use in each area, the percentage of seniors and children, and income level. After running the regression, she'd run a t-test on the coefficient of each variable to see which are statistically significant.

The null hypothesis for the test is that the coefficient is not markedly different from zero. A coefficient equal to 0 would indicate that there is no relationship between the independent and dependent variable—that the percentage of seniors in an area has no influence on the number of medical emergency calls, for example.

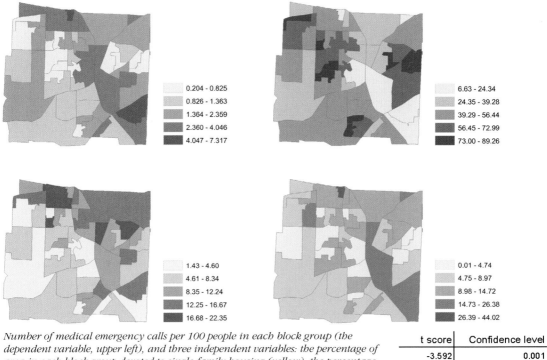

Number of medical emergency calls per 100 people in each block group (the dependent variable, upper left), and three independent variables: the percentage of area in each block group devoted to single-family housing (yellow); the percentage of the population that is age 65 and over (orange); and the percentage of each block group that is within a highway right-of-way (blue). The table shows the t-score and the confidence level for each independent variable. All three variables are significant at the 0.05 confidence level, or better, but the highway right-of-way variable is the most significant.

t score	Confidence level
-3.592	0.001
-2.554	0.014
6.213	0.000

You compare the test result to a critical value found in a table of t-test values. At a confidence level of 0.05, the critical value is 1.96 (assuming at least 120 degrees of freedom). If you get a value for the t-test higher than this, you reject the null hypothesis—you can be 95% sure you would not have gotten the predicted value if the coefficient were 0. The coefficient is statistically significant at the 0.05 confidence level. (See "Testing statistical significance.")

Even if a variable isn't significant, it may still belong in the analysis, if theory supports it, such as square footage predicting house value. You're pretty sure house value is partly dependent on square footage, but that variable may not show up as significant in the analysis. After careful consideration, though, you may still want to include it.

Factors influencing the regression analysis results

Ordinary least squares regression is only effective if your data and model meet all of the assumptions required by this method:

- There is a linear relationship between the dependent (y) variable and each independent (x) variable.

- The residuals have a mean of 0 (the overestimates balance out the underestimates).

- The residuals have a constant variance at all locations in the study area. The statistical term for this is homoscedasticity. The opposite condition (heteroscedasticity) can occur if the residuals increase or decrease along with the values of an independent variable. If heteroscedasticity is present, when you plot the residuals on the y-axis of a chart and an independent variable on the x-axis, the plot will appear cone-shaped.

- The residuals are randomly arranged along the regression line, rather than creating any pattern (high residuals grouped together, for example).

- The residuals are normally distributed—you can see this if you create a frequency curve.

- Independent variables are not highly correlated. If two or more independent variables are highly correlated (so-called multicollinearity), you get unreliable coefficient estimates and large standard errors (lots of variation in results from one sample to the next). To check for multicollinearity, run a regression analysis for each independent variable against all other independent variables and see if you get any very high r^2 values (close to 1). If so, you should drop at least one of the correlated variables. Since those variables are redundant, you only need one of them in your model.

Spatial data rarely meets all of these assumptions. You shouldn't apply ordinary least squares regression to spatial data without considering the assumptions of this method, and then modifying your model, data, or method whenever the assumptions cannot be met. Or use another method, such as the econometric spatial autoregressive approach, covered by Luc Anselin in his book *Spatial Econometrics: Methods and Models*.

REGRESSION ANALYSIS AND GEOGRAPHIC DATA

Even if your analysis meets all the assumptions of ordinary least squares regression, it may not be valid if it's not complete. Incomplete models are said to be misspecified. For geographic data, misspecification can result from several sources.

The process you're analyzing might be quite different in different parts of your study area. For example, when analyzing property values for a county, it may be that the factors contributing to property values in rural and urban portions of the county are quite different. To understand the process, you'd need to analyze rural and urban areas separately, or take the differences into account in your model.

Misspecification can occur if you're analyzing data at the wrong scale for the process. Perhaps you're using data collected for census tracts but you're asking questions that have nothing to do with census boundaries, like questions about air pollution. Or you're analyzing data summarized by county boundaries, but the actual geographic phenomena vary dramatically within each county. In these cases your data will be structured by the administrative units and so will your results.

You may also be missing variables. Suppose you want to predict temperatures in the United States. You could start by using latitude as your independent variable. This single variable, however, does not constitute a fully specified model, since it doesn't take into account seasonal variation or elevation. Wisconsin is hot in August even though it's in the northern latitudes, while the top of Mt. San Jacinto in southern California will be cold in February. If you used only latitude, your predictions would be wrong most of the time. Adding an elevation variable and a variable to depict seasonal variation would be a good next step. Your model results will improve, but if you check your predictions for Death Valley and for a point 260 miles due west on the coast (same latitude, same elevation, same season) you will see more evidence of misspecification. You would then add a "distance from coast" variable. Tracking down and including all the necessary independent variables can be difficult—especially if the process you're analyzing is not well understood.

As Robert Haining notes in *Spatial Analysis and GIS,* any systematic pattern in your residuals, including regional variation or spatial autocorrelation, is evidence that you are missing independent variables. This is why examination of residuals is such a large part of validating your model. The residuals represent everything you have not accounted for, or cannot, although even in a fully specified model there might still be small residuals that reflect random noise. As the processes you're analyzing get more complex, modeling them becomes a much bigger challenge. There will be times when you can't identify variables that will account for regional variation or spatial autocorrelation. So, statisticians have developed methods that attempt to account for these effects.

Dealing with regional variation
One approach for incorporating regional variation in your model is known as geographically weighted regression (GWR). Research on this method has been lead by Stewart Fotheringham, Chris Brunsdon, and Martin Charlton.

Geographically weighted regression works by allowing model coefficients to vary regionally. Essentially, you run a regression for each location, rather than for the study area as a whole. The equation for geographically weighted regression indicates that the predicted values and coefficients are for a single geographic location *(g)*.

A geographically weighted regression—the predicted values and coefficients are calculated for each location (g)

$$\hat{y}(g) = a + b_1(g)x_1 + b_2(g)x_2 + b_3(g)x_3$$

Each location has its own set of coefficients, so an r^2 value can be calculated for each location. You can then map the coefficients and r^2 values to get a sense of the relationship between the dependent and independent variables across the study area. If you were predicting in which counties low birth weights are likely to occur, and you know that your independent variables—per capita income and the percentage of the population living in rural areas, for example—vary across the study area, you'd use geographically weighted regression.

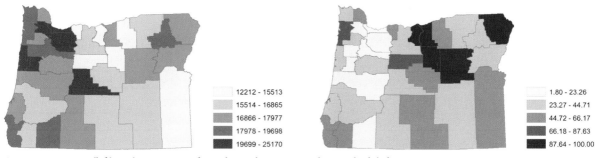

Per capita income (left) and percentage of population living in rural areas (right), by county

COUNTY	% LOW BIRTH WEIGHT	PCI	% RURAL	C PCI	T PCI	C % RURAL	T % RURAL	R-SQUARED
41009	3.63	17977	44.96	-0.00010	-0.488	-0.03147	-1.700	0.516
41007	4.74	18486	41.45	-0.00009	-0.426	-0.02872	-1.508	0.499
41059	6.36	16371	31.43	-0.00016	-0.788	-0.04908	-2.764	0.641
41063	0	17288	100.51	-0.00016	-0.716	-0.05400	-2.911	0.671
41049	5	15454	47.37	-0.00015	-0.789	-0.04600	-2.632	0.620
41061	5.50	18413	41.41	-0.00016	-0.763	-0.05156	-2.850	0.655
41021	5.56	17270	90.42	-0.00015	-0.771	-0.04360	-2.510	0.602
41057	5.49	19698	74.84	-0.00009	-0.457	-0.02890	-1.531	0.494

An r² value is calculated for each county, along with a coefficient and associated t score for each independent variable (in this case, per capita income and percent rural population).

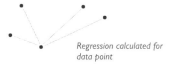

Regression calculated for data point

Regression calculated for sample location

The locations for which you calculate the regression depend on the type of data you're analyzing. For discrete features (points or areas), the locations can be the locations of the original data points, any location in the study area, or some combination of the two. As long as you have the geographic coordinates, you can calculate the r^2 values. You could then map the r^2 value for each point or area. For point locations representing spatially continuous data, you'd create a raster surface of the r^2 values. For contiguous areas you'd use the original features, since they cover the entire study area.

The coefficient for a location depends on the influence of the surrounding data points. The influence is based on how far the particular data point is from the location you're calculating the coefficient for—the closer the point, the greater the influence. The bandwidth—which you specify—defines the rate of distance decay; that is, the rate at which the influence decreases as the distance increases. The smaller the bandwidth, the more rapidly the influence drops off, and the greater the local variation in the coefficients.

Mapping the resulting r^2 values tells you where your model is working best (the higher the r^2, the better the fit). That's important if the purpose of your analysis is to predict values.

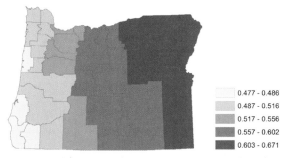

	0.477 - 0.486
	0.487 - 0.516
	0.517 - 0.556
	0.557 - 0.602
	0.603 - 0.671

The mapped r² values indicate the regression model produces a better fit in the northeast region of the study area.

Mapping the values of the coefficients shows you how each coefficient varies across the region, and where the variable has the biggest impact on the regression. A t-score is also calculated for each coefficient for each location. You can then map the locations where the t-score exceeds the critical value for specific confidence levels (the 0.01 and 0.05 levels, for example). That would show you where the dependence is statistically significant for each independent variable.

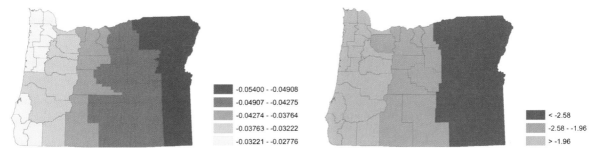

Mapping the coefficient for the percent rural variable shows how the relationship between low-weight births and the percent rural population varies across the study area, while mapping the t-scores for the variable (at the 95% and 99% confidence levels) shows where the relationship is statistically significant.

Dealing with local trends

Resampling and spatial filtering are two methods often used to address local trends. Resampling is appropriate when the local trends are causing redundancy and you want to remove the effects of spatial autocorrelation from your model. Suppose one of your variables is median rent for census block groups, and you find there is strong spatial autocorrelation—similar rent values cluster among nearby block groups. This suggests that the spatial unit you're using (block groups) is too small for that variable. Resampling to use a subset of block groups would potentially remove the spatial autocorrelation. Resampling is also often an effective strategy when modeling spatially continuous data represented as a raster.

Median rent by block group. Spatial autocorrelation is statistically significant at the 0.01 confidence level for this variable.

Spatial filtering is best when you want the spatial processes underlying the variable you're attempting to predict incorporated in your model. You theorize that the spatial processes are an important independent variable and that removing them would risk misspecification. For example, if you're trying to predict home prices and you suspect that the price of homes affects the price of other homes nearby, you'd want to capture that effect in your model—to help explain the process—rather than eliminating it.

Resampling
Resampling involves selecting a subset of data points on which to run the standard ordinary least squares regression.

One way to do this is to select sample locations that minimize spatial dependency while representing all of the characteristics of the dataset, or as many as possible. You'd do this by looking at maps of the features, symbolized using the different variables, and picking a set of features that does not include multiple features that have similar values and are near each other. Since you have to consider all the variables, this approach will be difficult if your model has more than a few variables.

An alternative is to use a random process to select sample locations. However, using randomly selected data points will not ensure that your sample will be free of spatial dependency.

A third approach is to select a sample based on measurements of the spatial autocorrelation in your dataset. Most geographic data exhibits positive spatial autocorrelation (features near each other are more similar than features farther away). The idea is to use a distance for which spatial autocorrelation is no longer statistically significant to set up your sampling scheme. If you find that spatial autocorrelation is no longer significant at four miles, for example, you can overlay your study area with a four-mile-by-four-mile grid, and then randomly select a location within each grid cell for your sample.

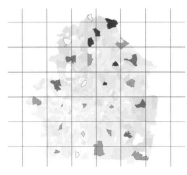

Spatial autocorrelation for median rent in this region is most significant at about four miles. By selecting a subset of block groups at this distance, using a four-mile-by-four-mile grid, spatial autocorrelation is reduced so it is no longer statistically significant.

To find the distance where spatial autocorrelation is no longer statistically significant, use a weighted K-function routine. Or run a statistic that identifies spatial autocorrelation (like global Moran's I). You'd run the statistic multiple times, using increasing distance bands (for example, one mile, two miles, three miles, and so on). The minimum distance should ensure that each feature has at least one neighbor and the maximum distance should ensure that no feature has all other features as a neighbor. (See chapter 3, "Identifying patterns.")

Once you have your sample, run a statistic that identifies spatial autocorrelation, such as global Moran's I, to ensure that none exists. If there is none, you can then run the standard ordinary least squares regression analysis on the set of sample points. Otherwise, you may need to choose a different sample, perhaps using a larger distance.

Spatial filtering
The idea behind spatial filtering is to figure out what component of each value of a variable is due to spatial autocorrelation. Each value is split into its spatial and nonspatial parts—the spatial component is filtered out of the variable. The spatial and nonspatial components each become variables in the regression analysis.

The assumption is that rather than being a factor that is biasing the analysis, any spatial autocorrelation can be captured and used as an additional independent variable to refine the model. The method has been pioneered by researchers Art Getis and Daniel Griffith.

One spatial filtering approach uses the Gi statistic (discussed in chapter 4, "Identifying clusters"). Gi requires a threshold distance—in this case, the distance at which spatial autocorrelation is strongest. As with resampling, you could use the K-function for weighted data, or run global Moran's I at a range of distances to identify the distance.

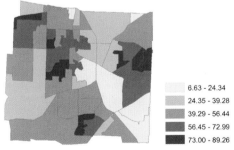

	6.63 - 24.34
	24.35 - 39.28
	39.29 - 56.44
	56.45 - 72.99
	73.00 - 89.26

Percent of each block group consisting of single-family housing. Spatial autocorrelation for this variable is statistically significant at the 0.05 confidence level, and is strongest at a threshold distance of about 4,000 feet.

Once you have the threshold distance, you filter each observation into spatial and nonspatial components. The equation compares the observed value of $G_i(d)$ with its expected value.

$$x^f_i = x_i \; \frac{W_i \, / \, (n-1)}{G_i(d)}$$

The difference between x^f_i and x_i represents the spatial component of the variable, x^{sp}_i.

$$x^{sp}_i = x_i - x^f_i$$

The expected value represents what the filtered value (x^f_i) would be if there was no spatial autocorrelation at that location. The ratio would be 1, and the original observed value of x would remain the same. If there is spatial autocorrelation among high values of the variable, the observed Gi value will be larger than the expected value, and x^{sp}_i will be positive. The converse is true if there is spatial autocorrelation among low values of the variable.

BLOCK GROUP	% SFR	GI	EXPECTED GI	% SFR FILTERED	% SFR SPATIAL
410670317021	0.333	0.073	0.059	0.270	0.063
410670312002	0.422	0.183	0.196	0.453	-0.030
410670312001	0.183	0.027	0.039	0.264	-0.081
410670311002	0.308	0.053	0.078	0.456	-0.148
410670304021	0.484	0.119	0.118	0.477	0.007
410670304012	0.305	0.104	0.157	0.462	-0.156
410670311001	0.510	0.094	0.118	0.640	-0.130
410670304011	0.233	0.029	0.039	0.311	-0.078

After filtering, each block group has filtered and spatial component values for the percent single-family housing variable, calculated using the ratio of the expected to the observed Gi values.

Once you have x^f_i and x^{sp}_i, the coefficients can be calculated, and the new independent variables added to the model. The original regression model looked like this:

$$\hat{y}_i = a + bx_i$$

The model after spatial filtering would look like this:

$$\hat{y}_i = a + bx^f_i + bx^{sp}_i$$

The process would be the same for any additional independent variables for which you want to capture the spatial component. After creating the spatial component variables, you should test for multicollinearity by running a regression of each spatial variable against each other one. Often the spatial components created from different independent variables will be highly correlated—especially if the data was collected by the same spatial unit (census tract, or county). If there is multicollinearity, you may need to remove one or more of the spatial variables.

RUNNING A LINEAR REGRESSION ANALYSIS WITH GEOGRAPHIC DATA

1. Determine what it is you are trying to predict.
This is your dependent variable or predicted value ($\hat{y}$). For ordinary least squares regression, it should be interval data rather than a ratio, ordinal (rank), binary, or nominal (categorical) data.

2. Identify the key independent variables.
Identify the variables that might explain the variable you are trying to predict. Often the information is not available, so you may have to use surrogate variables. If you're predicting where chaparral will occur and you don't have solar radiation readings for your study area, you'd have to use slope and aspect variables instead to capture the effect of solar radiation.

3. Examine the distribution of your variables.
You can do this is by creating a scatter plot. If there are clear curvilinear relationships, you may need to use a method other than ordinary least squares. If you find outliers in any of the variables, consider performing the ordinary least squares regression with each outlier and then without to see how sensitive your results are to those values. You should also look for strong correlation among independent variables (multicollinearity) and remove any redundant variables from the model.

If your dependent variable and most of your independent variables come from data collected using the same spatial units, such as census block groups, you should test each independent variable for spatial auto-correlation. The danger is that you may end up modeling the structure imposed by the spatial units, rather than the true relationships between your dependent and independent variables. If there is spatial auto-correlation, consider using the spatial filtering method.

4. Run the ordinary least squares regression.
You specify the dependent and independent variables, and the software calculates the coefficients and r^2 value, as well as the predicted values and residuals. Depending on the software you're using, it may also run the t-test for each coefficient.

5. Examine the coefficients for each independent variable.
You should remove irrelevant variables—those that are not statistically significant, whose coefficient is not markedly different from zero, or whose presence is not critical to theory behind your model.

6. Examine the residuals.
Each residual will be associated with an observation—a geographic feature, in this case—so you can map them or run additional spatial statistical tools with them.

- Using a tool like global Moran's *I,* test the residuals for spatial auto-correlation. Any statistically significant spatial autocorrelation may be evidence of a missing variable in your model. If strong spatial auto-correlation exists, consider using the resampling or spatial filtering methods.

- Map the residuals and look for clues regarding missing variables. The overestimates and underestimates should create a random pattern. If they don't, look for trends that suggest missing variables. If you're pre-dicting precipitation and didn't include an elevation variable, you would see underestimates in higher elevations and overestimates in valleys. You might need a distance-from-coast variable if you notice consistent under- or overestimates along the coast.

- Plot the predicted *y* values against the residuals to make sure there is no evidence of heteroscedasticity.

- Calculate the mean for the residuals—it should be zero (overestimates and underestimates balance out).

- Create a frequency curve for the residuals to make sure they exhibit a normal distribution.

Problems with the residuals are evidence of misspecification. Continue to look for missing independent variables until the model is fully specified. Ideally, close examination of the residuals will identify these variables, and your final model will no longer violate any of the assumptions. If, how-ever, spatial autocorrelation remains, consider either resampling (for raster data) or using the spatial filtering method. If regional variation among the residuals remains or you see evidence of heteroscedasticity, consider using geographically weighted regression.

Once all the assumptions have been met, you can interpret your results and report your findings.

Anselin, Luc. *Spatial Econometrics: Methods and Models.* Kluwer, 1988. Anselin presents econometric approaches to spatial regression.

Earickson, Robert J., and John M. Harlin. *Geographic Measurement and Quantitative Analysis.* Macmillan, 1994. The text covers basic correlation and regression analysis, and includes a discussion of data transformations.

Ebdon, David. *Statistics in Geography.* Blackwell, 1985. Ebdon discusses Pearson's correlation coefficient and Spearman's rank correlation coefficient, in a geographic context, as well as simple linear regression.

Fotheringham, A. Stewart, Chris Brunsdon, and Martin Charlton. *Geographically Weighted Regression: The Analysis of Spatially Varying Relationships.* Wiley, 2002. Seminal work on geographically weighted regression. The book presents both the theory and the application of the method, and includes a discussion of using the GWR™ software package.

Getis, Arthur. "Spatial Filtering in a Regression Framework: Examples Using Data on Urban Crime, Regional Inequality, and Government Expenditure." In *New Directions in Spatial Econometrics,* edited by L. Anselin and R. Florax, 172–85. Springer-Verlag, 1995. Getis presents a spatial filtering process using the Gi statistic, and shows how the method can be used with crime and economic data.

Getis, Arthur, and Daniel Griffith. "Comparative Spatial Filtering in Regression Analysis." *Geographical Analysis* 34, no. 2 (2002). The paper discusses two spatial filtering approaches—the Getis approach, which uses Gi, and the Griffith eigenfunction decomposition approach—and compares the two in an example using economic data.

Griffith, Daniel A., and Carl G. Amrhein. *Statistical Analysis for Geographers.* Prentice Hall, 1991. The text includes chapters on simple linear regression and multiple regression.

Haining, Robert. "Designing Spatial Data Analysis Modules for Geographical Information Systems." In *Spatial Analysis and GIS,* edited by Stewart Fotheringham and Peter Rogerson, 45–63. Taylor & Francis, 1984. Haining presents the issues involved in using statistics—particularly regression analysis—with GIS data.

Hamilton, Lawrence C. *Regression with Graphics.* Brooks/Cole, 1992. Hamilton's text covers the basics of bivariate and multiple regression, as well as nonlinear regression models.

Johnston, Kevin. "Using Statistical Regression Analysis to Build Three Prototype GIS Wildlife Models." In *GIS/LIS '92 Proceedings* 1, 158–65. 1992. Johnston presents several applications of the resampling method in wildlife analysis.

Data credits

The following organizations and individuals provided GIS datasets used to create the examples throughout the book.

Air quality data is used with permission of the California Air Resources Board.

AIDS incidence data is used with permission of the California Department of Health Services, Office of AIDS.

Dengue fever outbreak data is used with permission of CHAD/Community Health, Christian Medical College. Copyright © 2001–2004 CHAD/Community Health, Christian Medical College, Vellore, India. The associated road and boundary data was provided by Jayanth Devasundaram, MD, MPH.

Crime data is used with permission of the City of Redlands Police Department, Redlands, California.

Wolverine data is used with permission of Defenders of Wildlife—Oregon Biodiversity Project.

Data from the Portland RLIS dataset is used with permission of Metro Regional Services, Data Resource Center, Portland, Oregon. For more information, contact:

Metro Regional Services
Data Resource Center
600 Northeast Grand Avenue
Portland, Oregon 97232
503-797-1742
drc@metro.dst.or.us

Toxic sites data is used with permission of the Oregon Department of Environmental Quality.

Birth weight data is used with permission of the Oregon Department of Human Services.

Oregon environmental and hydrography data is used with permission of the Oregon Geospatial Data Clearinghouse, Salem, Oregon.

Oregon election results data is used with permission of the Oregon Secretary of State, Elections Division.

Tundra peregrine falcon migration paths data is used with permission of Raptor Research Center, Boise State University. The study of these birds was supported by funds from the U.S. Department of Defense, Legacy Resource Management Program, and other contributors, and the efforts of numerous collaborators. To cite this data or make further use of it, please contact Mark Fuller, Raptor Research Center, Boise State University, Boise, Idaho 83725, USA; 208-385-4115; mfuller@eagle.idbsu.edu.

Emergency response data is used with permission of Tualatin Valley Fire and Rescue, Aloha, Oregon.

Demographic data, including median income and urban/rural population, is used with permission of the U.S. Census Bureau.

Earthquakes data is used with permission of the U.S. Geological Survey (USGS).

Storm tracks data is used with permission of Weather Services International, 4 Federal Street, Billerica, Massachusetts 01821, www.wsicorp.com.

Wildlife observations and migration data is used with permission of the Wyoming Game and Fish Department.

Wyoming environmental and hydrography data is used with permission of Wyoming Geographic Information Science Center.

The source for the Wyoming vertebrate species data is E. H. Merrill, T. W. Kohley, and M. E. Herdendorf. 1996. *Wyoming Terrestrial Vertebrate Species Atlas*. Wyoming Cooperative Fish and Wildlife Research Unit, University of Wyoming, Laramie.

The source for Wyoming state and national parks, historic sites, and schools data is the *Wyoming Digital Atlas* developed by the University of Wyoming's Department of Geography and Recreation.

Wyoming election results data is used with permission of the Wyoming Secretary of State's Office and can be found at http://soswy.state.wy.us/election/96elect/gs-race.htm.

Software credits

The following GIS and statistical software packages were used to create the analysis examples in this book.

ArcGIS 9. ESRI, 2004. The Spatial Statistics toolbox within ArcToolbox™ included as core functionality in ArcGIS 9 contains tool sets for measuring geographic distributions, analyzing patterns, and mapping clusters. The tools are open source, written as Python® scripts.

CrimeStat: A Spatial Statistics Program for the Analysis of Crime Incident Locations (v 2.0). Ned Levine & Associates and the National Institute of Justice, 2002. A stand-alone software package that includes routines for performing spatial pattern and cluster analysis.

GWR, v. 2.0. The University of Newcastle, 2002. A stand-alone software package for performing geographically weighted regression, developed by Stewart Fotheringham, Chris Brunsdon, and Martin Charlton.

Microsoft® Excel 2000. Copyright © 1985–1999, Microsoft Corporation. The ubiquitous spreadsheet software allows you to make a variety of charts and plots, and contains a number of basic statistical functions.

Statistical Analysis with ArcView GIS. Jay Lee and David W. S. Wong, Wiley, 2001. A set of ArcView 3.x projects, available for download with the purchase of Lee and Wong's book, contain Avenue™ scripts for performing statistical analysis with point, line, and area features.

The tables on the next two pages list the sources of the tools used to create the examples in this book. The list is not by any means exhaustive—either in terms of the functionality of the software packages listed, or in terms of the availability of the tools. Spatial statistics is an evolving field—methods and tools are continually being improved, and new ones developed. A good source for finding spatial statistics software and related information is the clearinghouse maintained by the Center for Spatially Integrated Social Science (CSISS) at www.csiss.org/clearinghouse.

	ArcGIS 9 Spatial Statistics Toolbox	CrimeStat 1.1	Statistical Analysis with ArcView GIS
Chapter 2 Measuring geographic distributions			
Mean center	X	X	X
Median center		X	X
Central feature	X		
Standard distance	X	X	X
Standard deviational ellipse	X	X	X
Directional mean	X	X	X
Chapter 3 Identifying patterns			
Quadrat analysis			X
Nearest neighbor index	X	X	X
K-function	X	X	
Join count statistic			X
Geary's c		X	X
Moran's I	X	X	X
General G-statistic	X		X
Chapter 4 Identifying clusters			
Nearest neighbor hierarchical clustering		X	
Local Moran's I	X	X	X
Gi^*	X		
Gi			X

	GWR 2.0	Microsoft Excel	Statistical Analysis with ArcView GIS
Chapter 5 Analyzing geographic relationships			
Pearson's correlation coefficient		X	X
Linear regression		X	X
Geographically weighted regression	X		

Index

adjacency-based neighborhoods, 118, 136–38, 140–41, 143–44, 166, 176

area features, 6–8, 138–41, 184–86, 197–98. *See also* contiguous areas; discrete features

attribute values
    center calculations, 28–38
    clusters, 163–80
    geographic relationships, 196–202
    pattern identification, 78–79, 104–32
    spatial distribution, 18
    spatial neighborhoods, 135
    standard deviational ellipse, 48
    standard distance, 41

binary data, 106, 109

binary weighting, 166, 170, 176

bivariate analysis, 197

bivariate regression, 212

Bonferroni correction, 174, 180

center calculations
    central feature, 27, 28, 35
    location issues, 35–38
    by location or attribute, 28–32
    mean center, 26, 28, 33–34
    median center, 27, 28, 34
    outliers, 36
    weight specifications, 30–32

central feature, 27, 28, 35, 232

centroids, 33, 42, 95, 138, 150, 184–85

Chi-square (X^2) test, 85–87, 103

circular variance, 59–60

clusters
    control group comparisons, 160–61
    discrete features, 152–62
    Local Geary's c, 166
    Gi* statistic, 175–80
    importance, 4, 148
    local statistics, 77–78
    Local Moran's I, 166–74
    nearest neighbor hierarchical clustering, 152–62
    similar values, 163–80

clusters *(continued)*
    spatial neighborhoods, 139–41
    statistical analysis, 149–51

confidence intervals, 154–57

confidence levels, 11–12, 64–65, 69, 75

contiguous areas, 7–8, 10–11, 30–31, 37, 41, 68, 79, 105–6, 136, 150, 186, 220

continuous features, 7–8. *See also* spatially continuous features

continuous values, 10–11, 13, 106–7

control group, 12, 92, 101–2, 160–61

correlation analysis, 203–9

covariance, 204

data errors, 19. *See also* error

degrees of freedom, 86, 155, 206, 209

dependent variables, 211–26

directional mean, 51–59, 232

direction measurements
    importance, 45–48
    line features, 51–60
    mean calculations, 54–57
    points and areas, 46–50
    variance, 59–60

discrete features, 6, 10, 35, 51, 78, 106, 145, 150, 152, 220

distance bands, 80–81, 97

distance-based neighborhoods, 136–40, 144–45, 166, 176

distance decay, 142

dummy variables, 212

edge effects, 96, 102, 166, 174, 180, 188–89

error
    data errors, 19
    geographic data, 189–90
    sampling errors, 63
    Type I/Type II errors, 69

exponential distance decay, 142–43

extent. *See* geographic extent

feature distribution
    center calculations, 26–38
    compactness, 39–44
    direction measurements, 45–61
    orientation measurements, 45–61
    pattern identification, 80–103
    sampling assumptions, 67–69
    statistical analysis, 4, 10, 22–25

frequency distributions, 14–16, 18–19, 226

Geary's c, 107, 108, 118–20, 124–26, 166, 201, 232

Geary's c_i. See Local Geary's c

General G-statistic, 107–8, 127–32, 232

geographically weighted regression (GWR), 219–21, 226, 233

geographic data, 6–10, 183–90, 218–26

geographic distributions
    center calculations, 26–38
    compactness, 39–44
    direction measurements, 45–61
    importance, 22–25
    orientation measurements, 45–61

geographic extent, 92, 95, 126, 187–89

geographic patterns: See pattern identification

geographic relationships
    correlation analysis, 203–9
    importance, 192–95
    statistical analysis, 196–202

geographic scale, 77, 81–82, 126, 140, 153, 186–87, 209, 218

geostatistics, 5

Gi statistic, 175, 223–24, 232

Gi* statistic, 175–80, 232

global statistics, 77–78, 165

heteroscedasticity, 217, 226

histograms, 13–14

homoscedasticity, 217

independent variables, 211–26

interval data, 8–10, 13, 30, 106–7, 150, 204, 206–7, 211–12, 225

inverse distance weight, 142, 144–45

join count statistic, 105–6, 108, 109–17, 232

K-function, 80–81, 97–103, 223, 232

Kolmogorov-Smirnov test, 84–87, 103

k-order nearest neighbor index, 93–95

$L(d)$ transformation, 99–101

linear distance decay, 142

linear regression analysis, 211–18, 225–26, 233

line features
    data representation, 6, 184, 186
    direction measurements, 51–60
    orientation measurements, 51–60
    pattern identification, 94
    spatial neighborhoods, 138–39
    statistical analysis, 184
    See also discrete features

Local Geary's c, 166

local G-statistic, 175–80. See also Gi statistic; Gi* statistic

Local Moran's I, 166–74, 232

local statistics, 77–78, 165

local trends. See spatial autocorrelation

mean, 15–17

mean center, 26, 28, 33–34, 232

mean direction, 54–57

mean orientation, 54–57

median, 15, 17

median center, 26–28, 34, 157, 232

misspecification, 218–19, 226

modeling techniques, 5. See also regression analysis

Moran's I_i. See Local Moran's I

Moran's I, 107, 108, 118, 121–26, 223, 226, 232

multicollinearity, 217, 225

multinomial data, 106

multivariate analysis, 197

multivariate regression, 215

nearest neighbor hierarchical clustering, 152–62, 232

nearest neighbor index, 80, 88–96, 103, 232

neighborhoods. *See* spatial neighborhoods

nominal (categorical) data, 8, 105–6, 108, 186, 212, 225

normal distributions, 14, 15, 17, 65, 68, 90, 206, 226

normalization sampling, 68–69, 114, 124–25, 170

null hypothesis
 Local Moran's I, 170
 nearest neighbor index, 90
 pattern identification, 75
 Pearson's correlation coefficient, 205–6
 regression analysis, 216–17
 significance tests, 11–12, 64–65, 67–69
 Spearman's rank correlation coefficient, 209
 statistical analysis, 6

ordinal (ranked) data, 9, 186, 204, 206–8, 225

ordinary least squares regression analysis, 211–14, 217–18, 225–26

orientation measurements
 importance, 45–46
 line features, 51–60
 mean calculations, 54–57
 points and areas, 46–50
 variance, 59–60

outliers, 10, 19–20, 36, 50, 165, 174, 180, 225

pattern identification
 feature locations, 80–81
 feature values, 104–8
 Geary's c, 107–8, 118–20, 124–26
 General G-statistic, 127–32
 geographic scale, 77
 importance, 72–74
 join count statistic, 105–6, 109–17
 K-function, 80–81, 97–103, 223
 local statistics, 77–78
 Moran's I, 107–8, 121–26
 nearest neighbor index, 80, 88–96, 103
 quadrat analysis, 80, 81–87, 103
 statistical analysis, 75–79

Pearson's correlation coefficient, 204–8, 233

point features
 center calculations, 28–38
 clusters, 150, 152–62
 data representation, 186
 distribution measurements, 39–44

point features *(continued)*
 geographic relationships, 198, 220
 orientation measurements, 45–50
 pattern identification, 80–103
 spatial neighborhoods, 138–39
 spatial weights, 145
 See also discrete features

Poisson distribution, 14–15, 82

probability
 clusters, 149
 join count statistic, 105, 114–15
 nearest neighbor hierarchical clustering, 154–56
 pattern identification, 75
 quadrat analysis, 83
 significance tests, 11, 63–66

proportional weight, 143

quadrat analysis, 80, 81–87, 103, 232

r^2 values, 214–15, 217, 219–20, 226

randomization sampling, 68–69, 114, 124–25, 170

random patterns, 68, 72, 105, 114, 226

random samples, 64, 66

raster data, 80, 136, 139, 141, 179, 185–87, 196–98, 220, 221, 226

ratios, 10, 13, 30, 32, 106–7, 150, 186, 197, 204, 207, 211–12, 225

regional trends, 12, 183, 200

regional variation, 200, 219–21, 226

regression analysis, 211–26

resampling, 221, 222–23, 226

residuals, 214, 217, 219, 226

Ripley's K-statistic. *See* K-function

risk, 64, 69

row-standardized weighting, 144, 166, 170, 172

sample, 63–64, 66

sampling assumptions, 67–69. *See also* normalization sampling; randomization sampling

scale. *See* geographic scale

significance tests
 basic techniques, 11–12, 63–70
 Geary's c, 124–25
 General G-statistic, 130–31

significance tests *(continued)*
    Gi* statistic, 178
    join count statistic, 114–16
    K-function, 100
    Local Moran's I, 170–74
    Moran's *I,* 124–25
    nearest neighbor index, 90–91
    quadrat analysis, 84–87

spatial autocorrelation, 105, 165, 200–201, 219, 221–26

spatial filtering, 221–22, 223–26

spatially continuous features, 7, 10, 68, 105–6, 150, 186, 220, 221. *See also* contiguous areas; raster data

spatial modeling, 5

spatial neighborhoods, 135–45

spatial statistics
    applications, 4–5
    geographic data, 183–90
    importance, 2–3
    methodology, 6–12
    significance tests, 63–70

spatial weights, 135–37, 142–45. *See also* weighted values

spatial weights matrix, 136–37, 142–45

Spearman's rank correlation coefficient, 204, 206–9

standard deviation, 17, 65

standard deviational ellipse, 44, 46–50, 158, 232

standard distance, 39–44, 232

standard error, 90–91, 155–56

study area boundaries, 77, 96, 102, 161, 187–89. *See also* geographic extent; edge effects

summarized data, 8, 10, 32, 80, 106, 150, 160

t-distribution, 155–56

three-dimensional (3D) mean center, 30, 33

t-test, 205–6, 209, 216–17, 221, 226

Type I/Type II errors, 69

univariate analysis, 197

variance, 16–17, 119, 121, 130, 214

weighted values
    binary weighting, 166, 176
    Geary's *c,* 119–20
    General G-statistic, 128–29
    geographically weighted regression (GWR), 219–21
    inverse distance weight, 142, 145
    Moran's Index (Moran's *I*), 121
    proportional weight, 143
    row-standardized weighting, 144, 166
    spatial distribution, 18
    spatial neighborhoods, 135–45
    weighted central feature, 35
    weighted mean center, 33–34
    weighted median center, 34
    weighted standard deviational ellipse, 49
    weighted standard distance, 41, 42

weights. *See* spatial weights

Z-score
    critical values, 65, 69
    Geary's *c,* 124–25
    General G-statistic, 130–31
    Gi* statistic, 178–80
    join count statistic, 114–16
    Local Moran's I, 170–74
    Moran's *I* (Moran's Index), 124–25
    nearest neighbor index, 90–91, 103
    sampling assumptions, 67